LONG-TERM GROUNDWATER MONITORING DESIGN
THE STATE OF THE ART

SPONSORED BY
Task Committee on the State of the Art in Long-Term Groundwater
Monitoring Design of the Environmental and Water Resources Institute

American Society
of Civil Engineers
1801 ALEXANDER BELL DRIVE
RESTON, VIRGINIA 20191–4400

Abstract: This report contains a valuable summary of the state-of-the-art in groundwater monitoring network design and was prepared with the needs of analysts and practitioners in mind. Methods for the design of groundwater monitoring networks are reviewed. The methods include analytical and quantitative procedures that identify optimal sampling locations and frequencies to obtain representative groundwater quality data essential in the development of groundwater management strategies. The report also contains chapters on the objectives of long-term groundwater monitoring, data requirements for groundwater monitoring network design, case studies of long-term monitoring network design at field sites, and future research and technology transfer needs for this important field of groundwater hydrology and hydrogeology.

Library of Congress Cataloging-in-Publication Data

Long-term groundwater monitoring design : the state of the art / sponsored by Task
    Committee on the State of the Art in Long-Term Groundwater Monitoring Design of the
    Environmental and Water Resources Institute.
        p. cm.
    Includes bibliographical references and index.
    ISBN 0-7844-0678-2
        1. Groundwater--Quality--Measurement. I. Environmental and Water Resources
    Institute (U.S.). Task Committee on the State of the Art in Long-Term Groundwater
    Monitoring Design.

    TD426.L66 2003
    628.1'61--dc21   2003045245

Foreword

The American Society of Civil Engineers (ASCE) Task Committee on the State-of-the-Art in Long-Term Groundwater Monitoring Design prepared this report. The committee was formed in October 2000 by the Groundwater Management Technical Committee, which is part of ASCE's Water Resources Planning and Management Council of the Environmental and Water Resources Institute. The committee's work is a volunteer effort carried out by its 25 members from academia, industry, and government. The committee officers are Barbara Minsker, University of Illinois at Urbana-Champaign (chair); Ira May, Army Environmental Center (vice-chair); Donna Rizzo, Subterranean Research, Inc. (vice-chair); Gustavious Williams, Argonne National Laboratory (vice-chair); and Hugo Loaiciga, University of California at Santa Barbara (secretary).

The committee objectives are to (1) facilitate communication between researchers and practitioners involved in long-term monitoring design, (2) disseminate state-of-the-art methods for long-term monitoring design, and (3) identify and disseminate needs for future research and technology transfer in order to improve long-term monitoring design and implementation. To fulfill these objectives, the committee convenes special sessions and panel discussions on long-term monitoring design at professional conferences and has prepared this report on the state of the art in long-term monitoring design. For example, the committee sponsored sessions and a panel discussion at the World Water and Environmental Resources Congress, in Orlando, Florida, in May 2001, and a session at the American Geophysical Union Meeting in Washington, DC, in May 2002. It plans to sponsor further sessions and technology transfer activities, which will be announced at the committee website (https://web.ead.anl.gov/asce).

The preparation of this report was a joint effort involving many of the committee members and others. Primary authors of each chapter are as follows.

Chapter 1: Barbara Minsker, University of Illinois at Urbana-Champaign

Chapter 2: Gustavious Williams, Argonne National Laboratory

Chapter 3: Maureen Ridley, Lawrence Livermore National Laboratory, and Carol Stoker, Lawrence Livermore National Laboratory

Chapter 4: Hugo Loaiciga, University of California at Santa Barbara; Paul Hudak, University of North Texas; and Barbara Minsker, University of Illinois at Urbana-Champaign

Chapter 5: Donna Rizzo, Subterranean Research, Inc.

Chapter 6: Barbara Minsker, University of Illinois at Urbana-Champaign; Gustavious Williams, Argonne National Laboratory; and David Dougherty, Subterranean Research, Inc.

Many of the ideas in Chapter 6 were generated during a round-table discussion held at Argonne National Laboratory and during a panel discussion held at the World Water and Environmental Resources Congress in Orlando, Florida. Panel members at these events are as follows.

Long Term Monitoring Round-Table Discussion

Argonne National Laboratory, Argonne, Illinois
March 5, 2001
Chairs:
>Jack Ditmars – Argonne National Laboratory
>Barbara Minsker – University of Illinois at Urbana-Champaign

Panel Members (in alphabetical order):
>Kirk Cameron – MacStat Consulting, Ltd.
>Philip Hunter – U.S. Air Force Center for Environmental Excellence
>Ira May – U.S. Army Environmental Center
>Beth Moore – DOE Environmental Management Science Program
>Maureen Ridley – Lawrence Livermore Laboratory (by teleconference)
>Donna Rizzo – Subterranean Research, Inc.
>Gustavious Williams – Argonne National Laboratory
>David Wilson – U.S. EPA Region V

Panel Discussion on Long-Term Monitoring Issues and Needs

World Water and Environmental Resources Congress, Orlando, FL
May 21, 2001
Moderator:
>Barbara Minsker, University of Illinois Urbana-Champaign

Panel Members (in alphabetical order):
>Hugo Loaiciga, University of California at Santa Barbara
>Maureen Ridley, Lawrence Livermore National Laboratory
>Donna Rizzo, Subterranean Research, Inc.
>Gustavious Williams, Argonne National Laboratory
>David Wilson, U.S. EPA Region V

The following reviewers are acknowledged for their contributions to this document.

Charles Davis, PAI Corporation
David Dougherty, Subterranean Research, Inc.
Robert Gibbons, University of Illinois at Chicago
Paul Hudak, University of North Texas
Beth Moore, DOE Environmental Management Science Program
Patrick Reed, Pennsylvania State University
David Wilson, U.S. EPA Region V
Kathy Yager, U.S. EPA Technology Innovation Office

Finally, Leon Jeter is gratefully acknowledged for his extensive and careful editing of this document.

In alphabetical order, the following list gives the contact information for each of the committee members.

Beckmann, Dennis
BP Global Environmental
    Management BU
509 South Boston, Mail Code N-866
Tulsa, OK 74103
ph: 918/581-4817
beckmadd@bp.com

Cameron, Kirk
MacStat Consulting, Ltd.
10330 Mill Creek Ct.
Colorado Springs, CO 80908
ph: 719/532-0453
kcmacstat@qwest.net

Chan Hilton, Amy
Dept. of Civil & Environ. Eng.
Florida State Univ.
2525 Pottsdamer St.
Tallahassee, FL 32310
ph: 850/410-6121
abchan@eng.fsu.edu

Chang, Chi-Chung
Radian International
9801 Westheimer, Suite 500
Houston, TX 77042
ph: 713/789-9801
chichung_chang@urscorp.com

Davis, Charles
Professional Analysis, Inc.
P.O. Box 98518
Las Vegas, NV 89193
ph: 702/295-0541
davisc@nv.doe.gov

Ditmars, Jack
ANL
9700 S. Cass Ave.
Argonne, IL 60439
ph: 630/252-5953
jditmars@anl.gov

Dougherty, David
Subterranean Research, Inc.,
33 Enterprise Place, Suite 5,
Duxbury, MA 02332
ph: 781/710-1448
ddougher@subterra.com

Hill, Mary C.
U.S. Geological Survey
Water Resources Div.
3215 Marine St.
Boulder, CO 80303
ph: 303/541-3014
mchill@usgs.gov

Hudak, Paul
Dept. of Geography and Environ. Science
Univ. of North Texas
P.O. Box 305279
Denton, TX 76203
ph: 940/565-4312
hudak@unt.edu

Hunter, Philip
AFCEE/ERC
3207 North Rd.
Brooks AFB, TX 78235
ph: 210/536-5281
Philip.Hunter@hqafcee.brooks.af.mil

Gibbons, Robert
Biostatistics Lab at PI
Univ. of Illinois
Rooms 455-457 (MC 912)
1601 W. Taylor
Chicago, IL 60612
ph: 312/413-7755
rdgib@uic.edu

Johnson, Robert
ANL
9700 S. Cass Ave.
Argonne, IL 60439
ph: 630/252-7004
rlj@anl.gov

Lillys, Ted
HydroGeoLogic, Inc.
1155 Herndon Parkway, Suite 900
Herndon, VA 20170
ph: 703/478-5186
tpl@hgl.com

Loaiciga, Hugo (secretary)
Dept. of Geography
UC Santa Barbara
Santa Barbara, CA 93106
ph: 805/893-8053
hugo@geog.ucsb.edu

Maxwell, Reed
Lawrence Livermore National Laboratory
P.O. Box 808
Livermore, CA 94551
ph: 925/422-7436
maxwell5@llnl.gov

May, Ira (vice-chair)
U.S. Army Environmental Cntr.
Bldg. E4480
Aberdeen Proving Ground, MD 21010
ph: 410/436-6825
imay@aec.apgea.army.mil

McMullen, Scott
DOE Subsurface Contaminants
    Focus Area
Savannah River Site
P.O. Box A, Bldg. 703-A
Aiken, SC 29802
ph: 803/725-9596
scott.mcmullin@srs.gov

Minsker, Barbara (chair)
Dept. of Civil & Environ. Eng.
Univ. of Illinois
3230d Newmark
Urbana, IL 61801
ph: 217/333-9017
minsker@uiuc.edu

Moore, Beth
DOE/HQ/EM-52
Forrestal Building, Room 3E066
1000 Independence Ave. SW
Washington. DC 20585
ph: 202/586-6334
beth.moore@em.doe.gov

Newell, Charles
Groundwater Services Inc.
2211 Norfolk, Suite 100
Houston, TX 77098
ph: 713/522-6300
cjnewell@gsi-net.com

Pinder, George
Dept. of Civil & Environ. Eng.
Votey Hall
Univ. of Vermont
Burlington, VT 05405
ph: 802/656-8697
george.pinder@uvm.edu

Reed, Patrick
Dept. of Civil & Environ. Eng.
215B Sackett Building,
Pennsylvania State University
University Park, PA 16802
ph: 814/863-2940
preed@engr.psu.edu

Ridley, Maureen
LLNL, L-528
P.O. Box 808
Livermore, CA 94551
ph: 925/422-3593
ridley1@llnl.gov

Rizzo, Donna (vice-chair)
Subterranean Research Inc.
P.O. Box 1121
Burlington, VT 05402
ph: 802/658-8878
drizzo@subterra.com

Shoemaker, Christine
School of Civil & Environ. Eng.
Hollister Hall
Cornell University
Ithaca, NY 14850
ph: 607/255-9233
christine_shoemaker@cornell.edu

Tuckfield, R. Cary
Westinghouse, Savannah River Co.
Building 773-42A
Aiken, SC 29808
ph: 803/725-8215
cary.tuckfield@srs.gov

Valocchi, Albert
Dept. of Civil & Environ. Eng.
Univ. of Illinois
2527c Hydrosystems
Urbana, IL 61801
ph: 217/333-3176
valocchi@uiuc.edu

Williams, Gustavious (vice-chair)
ANL
9700 S. Cass Ave.
Argonne, IL 60439
ph: 630/252-4609
gpwilliams@anl.gov

Wilson, David
U.S. EPA Region V
77 W. Jackson Blvd.
Chicago, IL 60604
ph: 312/886-1476
wilson.david@epa.gov

Yager, Kathleen
U.S. EPA Technology Innovation Office
EPA-ERT Bldg. 18, MS101
2890 Woodbridge Ave.
Edison, NJ 08837
ph: 732/321-6738
yager.kathleen@epamail.epa.gov

Zillmer, Doug
Naval Facilities Engineering
    Service Center
Code ESC413
1100 23rd Avenue
Port Hueneme, CA 93043
ph: 804/982-1556
zillmerda@nfesc.navy.mil

Contents

Chapter 1: Executive Summaries

1.1. Executive Summaries

Chapter 1 provides an executive summary of each of the succeeding five chapters included in this report.

1.2. Chapter 2: Introduction

[handwritten margin note: Collection of data for documenting LTM] Changes in gw or gw contamination at envir. restoration sites

Chapter 2 is an overview of long-term monitoring (LTM) issues and explains the scope of this document, which encompasses LTM design for environmental restoration sites. Herein, the acronym LTM refers to the systematic collection of data for documenting changes in groundwater or groundwater contamination at environmental restoration sites. Although many of the methods and issues presented may be useful for other applications (such as for waste storage facilities) and other media (such as surface water), they are not our focus.

The preparation of this report was motivated by the growing need for, and costs of, LTM as sites complete their active remediation efforts. Groundwater contamination still exists and is likely to remain for a long time at many of these sites. Long-term monitoring is essential at these sites to ensure the protection of human health and the environment and to monitor ongoing changes in the subsurface. Long-term monitoring is also essential for obtaining stakeholder acceptance when active remediation ceases yet contamination remains. The growing need for LTM is accompanied by an increasing realization that LTM costs will become significant in the foreseeable future. This need has led to a substantial interest in optimizing LTM designs in order to reduce long-term costs while protecting human health and the environment.

Our purpose is to disseminate methods that can be used for optimizing LTM designs and to provide case studies of field applications of these methods. Future needs for optimal LTM design are also discussed, both to provide guidance for research and policy efforts and to help practitioners identify areas where methods are still changing and improving. This document is meant to be a guide, with references to the primary sources for the detailed information, and not as an in-depth tutorial. It primarily focuses on the physical design of monitoring networks, rather than on the merits of various sampling methods (e.g., remote sensing methods or indicator samples vs. traditional well-based samples) or on the various goals or strategies for LTM.

As we use the phrase, the physical design of monitoring networks involves identifying which constituents and variables should be measured and their locations and frequencies, the data collection/validation/analysis procedures to be followed, the decision rules used for determining whether a problem exists, and the decision rules for assessing whether the underlying model or perhaps even the program itself needs maintenance. Most of the document concentrates on monitoring locations and frequencies for a single constituent or parameter because the current state of the art primarily addresses those aspects of LTM design. (See Chapter 6 for suggestions on

[handwritten margin note: identify coc, locations and frequencies, and data collection/validation/analysis procedures, decision rules (whether a prob exist), assessing whether the model or program need maintenance]

1

research needed to expand the state of the art so as to address the remaining aspects of LTM design.) An optimal design should minimize the number of sampling locations and frequencies while meeting the objectives of the LTM system. One of the most important steps in optimizing LTM designs is to identify the objectives of the LTM system. Many objectives are possible, ranging from broad, qualitative objectives, such as "protecting human health and the environment" to focused objectives, such as "verifying flow containment." Monitoring design usually centers on achieving clearer objectives that can be assessed in quantitative terms, as surrogates for the broad, qualitative objectives. A detailed list of the objectives that have typically been used for LTM design in the past can be found in Section 2.5.3.

1.3. Chapter 3: Environmental Data Management

Chapter 3 provides an overview of data management, an important issue in long-term monitoring. Data management is the process of understanding the data needs of an organization and making the data available to support the organization's operations. The ultimate goal of data management is to provide the seamless access to, and fusion of, massive amounts of data, information, and knowledge within a heterogeneous and real-time environment, as well as to support the functions and decision-making processes of an organization. Questions that must be asked in the pursuit of proper data management are, who is going to use the data, what types of data need to be stored, and how will these data be accessed? Based on the answers, a data management system can be created or an existing system can be modified to meet the needs of the organization.

The essence of a good data management system is providing the end-user with a consistent data set of known quality. The system should be one that is designed based on how the data are collected and processed, is very well documented, has specifically defined data elements, and has supporting data documentation. The supporting documentation includes quality control data that are carried along with the analytical data and other meta-data (information about the data, such as summary statistics, error estimates, analytical methods, and quality assurance/control results and flags).

Data sets get better the more they are reviewed. As errors and inconsistencies are identified and corrected, the data set improves. A good data management system will achieve this goal because its design promotes use, and the ultimate value of data is in its use rather than its storage. The development of Internet access tools and existing environmental data management systems can reduce the effort and cost associated with setting up a data management system.

1.4. Chapter 4: Methods for Optimizing Long-Term Monitoring Designs

Chapter 4 provides an overview of various methods for optimizing LTM designs. Most of the early work in monitoring design focused on methods for siting new monitoring wells. At that time, the focus was on site and plume characterization or plume detection. Most of these methods can also be used for LTM design. Recently, methods have been developed specifically for identifying sampling plans

2

that minimize spatial and temporal redundancy in existing monitoring wells. These methods are introduced, with the primary references provided for the interested reader to obtain details. More detail is given for methods for which accessible primary references are not available.

The methods summarized in this chapter are organized according to the types of tools used and the level of complexity involved. In selecting an appropriate method for LTM design, both the monitoring objectives and the amount of data and information currently available should be considered. Guidance is provided for selecting methods that are appropriate for different objectives and levels of data and information.

1.5. Chapter 5: Field Studies of Long-Term Monitoring Design

Chapter 5 summarizes nine recent field studies in which LTM design methods have been applied, along with references for previous LTM design field studies. The field studies illustrate several important findings for LTM design, including the relationship between site remedies and performance monitoring and the importance of considering anticipated contaminant plume shrinkage rather than, or in addition to, plume stability in selecting design methods.

Although regulatory and permitting processes usually require that site remedies be approved in a record of decision prior to the initiation of a monitoring plan, the remedy and the monitoring are inextricably related. The performance of the remedy can be evaluated only by monitoring for parameters that indicate whether its performance objective is being met. As data continue to be collected and site knowledge increases, the selected remedy may change, in which case so should the corresponding performance-monitoring scheme. Long-term monitoring plans must be designed to remain compatible with the selected remedies.

Finally, the importance of plume shrinkage vs. plume stability is highlighted; seven of the nine field studies presented in this chapter involve that objective. The distinction between verifying plume shrinkage and plume stability is particularly important for selecting an appropriate LTM design method. In the first case, the plume is (anticipated to be) moving, and thus complex nonstationary design methods may be needed. In the second case, the rate of plume change is small and simpler methods can be used. In either case, the amount of historical data available for LTM design influences which analysis methods are appropriate, not simply which ones can be used.

1.6. Chapter 6: Future Needs for Long-Term Monitoring

Chapter 6 presents the committee's recommendations for further research and new implementation guidance needed to improve the state of the art in LTM design. References to other published recommendations, where available, are also summarized. The recommendations are organized into two broad categories: research needs and technology transfer needs.

1.6.1. Research Needs. As sites move toward long-term monitoring, monitoring objectives shift from site characterization to performance assessment. Performance assessment involves collecting data to identify whether the remediation is progressing as expected. More accurate and cost-effective performance assessment is needed, which will require research to (1) develop methods for integrating and using nontraditional data, such as sensor or screening technologies in dynamic field sampling approaches; (2) identify and characterize variability in long-term monitoring data; (3) improve decision rules and protocols for long-term management; (4) develop "living" performance assessment models that are updated, refined, and used to analyze current conditions at the site and to predict and accommodate the effects of observed changes over the long term, (5) improve methods for linking remediation processes and LTM, (6) investigate methods for designing LTM systems at sites with fractured bedrock or Karst terrain, (7) develop improved methods for failure analysis and robust LTM design, (8) improve electronic management of and access to data, and (9) create publicly available data archives and test data sets.

1.6.2. Technology Transfer Needs. The committee strongly encourages the development of uniform LTM guidance, regulation, and practice across regulatory jurisdictions. To achieve this goal, the following technology transfer developments are needed: (1) regulatory guidance on the amount and types of data needed to demonstrate system integrity or performance and to identify appropriate LTM plans, (2) regulatory guidance and acceptance of dynamic sampling (or operation) plans, (3) regulatory guidance on incorporating LTM needs early in the remedial process, and (4) professional guidance and education on the special problems associated with LTM and the methods available for LTM design, using mechanisms such as workshops and regulator guidance.

Chapter 2: Introduction

2.1. The Need for Long-Term Monitoring at Environmental Restoration Sites

Long-term monitoring (LTM) of groundwater at restoration sites is key to maintaining and improving the quality of groundwater resources for the future. Long-term monitoring refers to the systematic collection of groundwater data for determining physical, chemical, and biological characteristics across time scales deemed adequate to demonstrate the achievement of water-quality goals or to document long-term changes in groundwater or groundwater contamination. Among the potential objectives for LTM are information gathering for the performance assessment of remediation systems, compliance assessment at remediated sites, risk assessment of source aquifers, and the determination of ambient groundwater quality or background parameters to use as a point of comparison for LTM systems. Long-term monitoring systems can be integrated into an environmental decision framework to help make future decisions about cleanup activities or needs and about future monitoring requirements.

This report specifically addresses the monitoring of environmental restoration sites. Although there are other situations where LTM might be needed or appropriate (e.g., for waste storage facilities or other engineered containment structures), these are not the focus of this report. Even so, many of the design techniques and issues presented here may be useful when addressing other LTM uses. The need for groundwater LTM at environmental restoration sites is growing as sites complete remediation efforts of surface soils and other sources of groundwater contamination. At many of these sites, groundwater contamination persists and it is not practical to restore the groundwater quickly to an uncontaminated state. Groundwater LTM at such sites is required to ensure that there are no unacceptable risks to human health or to the environment from this remaining contamination.

When environmental remedial actions were first started in the United States, it was widely believed that when cleanup was completed, the site would be restored to uncontaminated or even pristine conditions and that no further action would be needed. As site restoration activities have progressed across the nation and we have gained additional experience, we have come to realize that in many cases it is impossible or infeasible to restore a site to a pristine condition with active cleanup efforts, especially for sites with groundwater contamination (EPA, 1993). At sites where complete restoration is impossible, we have changed the goal of restoration to be the protection of human health and the environment rather than restoration to an uncontaminated state. As cleanup sites reach the end of their active remedial efforts, there is a need to monitor the remaining groundwater contamination to provide assurances that cleanup objectives are met and that remaining contamination is not posing an unacceptable risk to human health or the environment. In addition, there is often a need to provide documentation that the cleanup activities were successful and that the site no longer poses an unacceptable risk.

Estimated costs for the remediation and monitoring of existing groundwater contamination are significant. The National Research Council estimated that the cost of remediating polluted groundwater in the United States is somewhere between half a billion and a trillion dollars (National Research Council, 1993). The U.S. Department of Energy (DOE) estimated, in a report to Congress that it will spend approximately $5.5 billion on long-term stewardship between 2000 and 2006 (DOE, 2001). The majority of these DOE estimated costs would be for the long-term monitoring of groundwater contamination. The report also estimates that long-term stewardship costs will be on the order of $100 million per year for the next 70 years. In addition, the report acknowledges that the estimated 100 million per year is low because of uncertainties in future costs and technologies and the fact that these estimates excluded some known DOE sites and any clean-up actions in the DOE complex taken after 2006 (DOE, 2001). The Department of Defense is also concerned about LTM. The U.S. Navy estimates that LTM and remedial action operation costs will increase from about $40 million per year in 2000 to nearly $80 million in 2004 (U.S. Navy, 2001). The Navy also estimates that LTM monitoring costs will be about 30% of the installation restoration budget by 2004.

2.2. The Definition of Long-Term Monitoring

Long-term monitoring is defined as the testing of groundwater over an extended time period in order to document groundwater conditions, including the collection of chemical data, such as pH or contaminant concentrations, and physical data, such as water levels or temperature and, in some cases, bacteria types or numbers. For most LTM programs, the length of operation is generally not defined in terms of time. Instead, the length of operation is defined by performance objectives; if certain conditions are met (or occur), system operation can be terminated. Thus, LTM programs can be modified or terminated at a point in the future in response to changed conditions determined by the LTM system itself or due to changes in the objectives that eliminate the need for further monitoring. Any groundwater-monitoring program that has an indefinite time frame, established monitoring objectives and/or procedures, and a way for changing those methods or procedures can be termed an LTM.

There are a number of other definitions of LTM in use today, some of which are agency specific and relate to funding sources available in a specific agency for establishing and operating LTM systems. In some cases, not all groundwater monitoring that occurs, even if over relatively long time-periods, is classified as LTM. For example, the U.S. Army deems LTM programs as starting once active restoration activities at a site are complete. Typically, before this point in time, a monitoring network is designed and installed and the monitoring begun as part of site restoration, but activities that occur during active restoration are not referred to as LTM even though these actions may occur over years or decades. In another example, the U.S. EPA deems that LTM begins once a remedy is in place and operating correctly, well before the termination of active restoration (EPA 1996). Under this definition, monitoring systems installed and used as part of a remedy, regulatory policy, stakeholder requirements, or public relations are considered to be LTMs whether or not any contamination is present.

Two features distinguish LTMs from other forms of monitoring. One is that subsurface conditions should be predictable and generally stable. These conditions can result from passive remediation systems (e.g., monitored natural attenuation of VOCs), an ongoing remediation system, a source control system (such as engineered subsurface barriers), or even the absence of contamination. The goal of these LTM installations is to verify that the system continues to behave as expected. The second feature is that, generally, an LTM program should function as efficiently and routinely as possible, with minimal demands on resources, because of the long time horizon of potential operation. This second feature becomes particularly important as other activities at a site wane, and it is a goal of optimal design.

Long-term monitoring can be required at a variety of environmental cleanup sites and especially restoration sites where groundwater contamination is still present at the termination of active restoration activities or at disposal sites were the objectives might be to verify the conditions of the containment system. The decision to implement an LTM can be made based on the need for information about restoration systems or activities, the need to provide assurances about subsurface conditions, or regulatory requirements.

Long-term monitoring installations can have a number of regulatory drivers, including the Resource Conservation and Recovery Act (RCRA) (42 CFR, part 82), the Comprehensive Environmental Response Compensation and Liability Act (CERCLA) as amended (42 CFR, part 103), and the Underground Storage Tank (UST) programs (42 CFR, part 103). Monitoring programs executed under regulatory oversight usually have regulatory-defined contaminants of concern (Cocks) and action levels (the level of contaminant concentrations that require action) for these Cocks. In addition, sampling frequencies and monitoring point locations may be stipulated. The regulatory environment for LTM is extensive and complex and varies considerably for each site. I intend this overview only to highlight the issue of regulatory drivers; it is not intended to provide an exhaustive review of regulations that might be applicable for LTM at a given site.

Long-term monitoring programs mandated by regulation will often have stock lists of COCs and action levels, but these should be treated as "defaults" to be used only until site-specific lists are developed. Decision Quality Objectives (rather than Data Quality Objectives) should be developed that are related to the specific circumstances of a site and take into consideration potential risks, exposure routes, and concerns. Monitoring parameter lists should be relevant to the contaminants that are likely present based on a prior knowledge of the site. Similarly, action levels (in principle) should be determined using site-specific contaminant transport considerations, sensitive receptor locations, and current and potential site uses. These factors also hold true for monitoring frequencies and locations. Monitoring could possibly involve more frequent measurements of short lists of indicator parameters that are augmented periodically by other COCs.

Programs implemented without regulatory oversight may be more flexible in operation, and it may be easier to adapt system objectives to meet changing conditions. However, no matter what drives the implementation, all LTMs should have a conceptual model of where and how contaminants are moving, or might be moving, and of whether the contaminants are undergoing, or could undergo, important chemical or physical changes, such as in situ degradation. Long-term monitoring network designs should be based on these concepts. One of the objectives of an LTM program is to validate the conceptual model and to identify when a system modification might be needed or possible (e.g., if the monitoring results do not match what is expected). To meet these objectives, LTM programs may track changes in areas with contaminated groundwater, monitor previously uncontaminated areas to provide assurance that they remain unaffected, or monitor uncontaminated areas to establish background conditions.

Long-term monitoring programs are not designed for site characterization, although much of the equipment installed and data collected for characterization can be used for LTM. Designing an LTM program includes establishing a set of monitoring parameters and COCs, although the COCs may not be measured directly by the LTM instruments. For example, indicator measurements, which are measurements that indicate the possible presence or changing concentration of a COC without measuring that parameter or value directly, might be used to demonstrate that such conditions exist that preclude the COCs from posing an unacceptable risk, rather than measuring the COCs themselves, or to indicate changes in the subsurface environment that may be deemed important. In an LTM program, there may be a predetermined set of circumstances in which the data gathered, the methods used, or the measurement frequency employed can be adjusted with changing conditions. Typically, over time, the number of locations observed, the number of parameters measured, and the measurement frequency are reduced as the LTM program provides more information to validate the site conceptual model and to document the success or failure of the remedy used for restoration (AFCEE, 1997).

In general, LTM programs can be thought of as providing a quality assurance function for closure or other remedial actions. Over time, if an LTM program gathers data that differ significantly from data that were expected, based on the conceptual model and design assumptions, the initial analysis might need to be revisited, including decisions made about the site cleanup and the monitoring program, adding or deleting sampling points or analytes, changing monitoring frequencies or detection limits, and so on. The LTM data can also be used to update the conceptual model of the site and determine whether other actions are needed or whether current efforts are appropriate. These additional actions can range from additional cleanup actions (if the data indicate an unacceptable risk to human health or the environment), to redefining the LTM objectives (if stakeholder objectives are different or have changed from the initial assumptions), to determining that the restoration activities and monitoring were successful and the LTM program can be terminated. Determining what actions and whether future actions are required are among of the main purposes of an LTM program.

2.3. Purpose of this report

2.3.1. Needs for Optimal LTM Design. This book focuses on the methods and tools used for the optimal design of an LTM program. The operation and management of LTM systems can be expensive. By definition, these systems operate for a long time and have recurring costs. Under these conditions, small cost savings or efficiencies can be magnified by repetitions of sampling events. The optimal design of LTM networks attempts to minimize system costs while meeting the objectives of the program. The tools and methods discussed and presented in this report can provide practitioners with ways to design and implement efficient LTM networks.

An optimal design can formalize the procedures by which an LTM system can be revised as new information and/or methods become available. An optimally designed LTM system continually generates new information that is evaluated to determine whether further efficiencies can be realized. This evaluation can determine whether the correct data are being collected in light of changed conditions; whether the LTM design and operation are still appropriate and meeting stated objectives; and determine whether some of the long-term objectives have been met or whether these objectives are still appropriate in view of changing site conditions, technology, regulations and other policies, and stakeholder expectation Typical objectives that have been used to design LTM systems are given in Section 2.5.3.

Finally, it should be noted that benefits from using optimal LTM network design are not necessarily only from the monetary savings. The benefits realized may also be greater reliability, more accuracy, or more acceptance of the LTM program on the part of stakeholders at a site. Failure to demonstrate adequate oversight may result in reopening a site to investigation at significant costs. The ability to reduce these future liabilities needs to be considered when evaluating the costs of an LTM program.

2.3.2.   New Methods Dissemination. Researchers and practitioners have recognized the need for efficient, effective methods for LTM design and have developed a number of new methods and tools to aid in the optimal design of an LTM program that will meet required objectives while minimizing costs. These methods and tools range from relatively simple procedures for analyzing site information to complex numerical methods that determine the minimum (or maximum) of a formalized objection function for a given set of constraints and may require significant amounts of data for effective implementation.

Unfortunately, existing optimal design methods are not in widespread use at restoration sites. We believe that these optimization methods are applicable for LTM design and are usable by the groundwater restoration community. We also believe that the use of these tools can result in substantial cost-savings, while providing a LTM program that will meet all design requirements.

In this light, one purpose of this book is to familiarize the reader with these new ideas and to disseminate these developing methods and tools for optimal LTM design to a wider audience. We provide the groundwater community with a general description of the numerous categories of tools available and give readers information about the applicability of tools from a given category.

Currently available optimization tools can produce an efficient LTM design, if they are correctly used. A barrier to using these optimal design tools is that many practitioners in the environmental restoration field are not familiar with the methods and their benefits or when the use of a specific tool or method would be appropriate. One of the goals of this book is to overcome this barrier by describing the types of tools that are available, when a specific tool might be applicable, and give examples of successful uses of these developing optimal design tools.

An additional purpose is to identify future needs for optimal LTM design research and development. By providing descriptions of the categories of tools available to the groundwater community and the situations where these tools are applicable, we identified areas in which additional tools and methods are needed. The last chapter of this report enumerates the needs we identified and can be used by the groundwater community to direct research efforts and to help practitioners identify areas in which particular tools or methods may still be changing and improving.

2.4. Document Scope

This document summarizes tools, methods, and case studies for the optimal design of an LTM network and provides references to primary sources for details on the methods that are presented. Historically, the main focus of optimal design has been on the physical portion of LTM networks: determining measurement locations and frequencies. Other LTM design parameters, such as measurement types, measurement costs, and data analysis methods, can also significantly affect design objectives but have not been substantially addressed in most currently available methods. A large number of techniques can be used for optimal LTM design, ranging from decision tree analysis, which can require few data or specialized tools, to approaches that employ sophisticated computer modeling to implement. In most cases, the optimal design of a LTM network will involve various combinations of these tools and methods, as demonstrated by the case studies.

As noted, LTM programs can have a number of objectives that range from items that can be technically defined, such as "are contaminant concentrations at a point below a specified level," to qualitative objectives, such as providing assurances to stakeholders that a site does not pose an unacceptable risk to the community. When designing an LTM program, the issues related to various site-specific objectives need to be considered, including such issues as risk receptor locations, future site uses, and the need to demonstrate remedy effectiveness, in addition to the issues involved in the physical design of the LTM. Although all these issues are important for the design of an effective LTM program, detailed discussions of these qualitative objectives are beyond the scope of this report.

This report focuses on current and emerging techniques for the optimal design of LTM networks and includes issues, such as the following:

- where to locate measurement points
- how often to measure at a given measurement point
- how to optimize measurement types and frequencies, incorporating such objectives as minimizing cost, minimizing risk, and minimizing error
- how to resolve data quality and data management issues related to the design and operation of these systems
- how to establish data storage/analysis techniques and procedures
- Discussions on future research and technology transfer needs

These physical design issues are complex and can be approached in numerous ways. In addition, it is not always practical to separate the sometimes-competing objectives when developing an optimal LTM program. A decision on one issue can significantly impact another. For example, if an LTM program is designed to use indicator samples, in addition to traditional sampling, the physical location of the sampling wells and the sampling frequency might be affected by this choice and could be different than the design of the same LTM program with only traditional samples. The developing nature of this field indicates that system design should be flexible to allow the incorporation of future advances in technology and understanding. The optimal design of an LTM program helps practitioners to consider and efficiently address these various issues and provides a framework for the continued refinement of an LTM program.

This report focuses on the tools and methods that can be used to optimally design the technical portion of an LTM program. In general, it is beyond the scope of this document to address many of the other important issues that may influence an optimal LTM program design beyond acknowledging their importance. Instead, where possible, these issues are noted and discussed briefly.

This document does not evaluate or recommend sensor types, sampling protocols, or analytical measurement techniques or look at other important issues, such as long-term well integrity and maintenance. These issues can have a large impact on LTM programs. For example, changing field-sampling procedures from bailing three to five well volumes before sampling to using micropurge techniques can result in different contaminant concentrations being reported, resulting in apparent system changes where none actually exist. Changing analytic procedures in the laboratory can have similar results. In addition, over time, changes commonly occur in the geochemistry, hydrology, and other physical conditions in the subsurface. For example, changes in measured chemical concentrations can be caused by variations in recharge due to changing surface vegetation or cover as easily as by the amount of chemical emanating from a potential source of concern. All these issues are important in the design, implementation, and operation of an LTM network.

Operating environmental monitoring systems over long-term periods will bring to light a number of issues that are not present with systems designed for short-term efforts. While noting that these issues exist, this report does not try to identify all the possible problems that may result during long-term operation of monitoring systems. For example, well integrity issues may become important in the future, although they have not been as significant during shorter monitoring programs. Currently, only a few sites have experience with sampling wells over long time periods and using the resulting data to analyze and confirm trends.

For long-term operation of LTM systems, conditions and activities on the surface above the contaminated groundwater can also have an impact. Wellhead protection will be an important concern for sites where wells will be in place and needed for several decades after active restoration has been completed. Establishing ways to protect these wells and ensuring that surface activities do not affect sampling results will be critical, although not addressed here.

2.5. Background on Optimal LTM Design

Optimal LTM design objectives can be based on technical requirements, such as information on how a remedial system is operating, or compliance requirements, such as assurance that previous uncontaminated areas remain uncontaminated. All LTM programs require data collection and analysis over a long time period; optimally designed LTM networks (and programs) should perform this function in the most efficient way possible. Cost and effort should be minimized consistent with providing a desired level of information as established by the site's Decision Quality Objectives.

The optimal design of LTM networks should result in a system that will:

- be protective of human health
- be protective of ecological health
- be cost effective
- provide assurances that placed remedies are adequate
- improve monitoring time horizons (provide for reductions as time progresses and, ultimately, a termination),
- provide sufficient and appropriate information on which to base future decisions (such as interventions or updates to the LTM program or its underlying conceptual model)
- provide estimates of uncertainty in the LTM assessment

Optimal LTM networks need to include future decision requirements as part of the initial design. This makes optimal LTM design an iterative process, in which collected information is used to determine and change future operations and configurations. Future changes can be predetermined based on expected monitoring results (e.g., a decision rule that reads "if the contaminant level in well X exceeds 10 ppb, new wells will be installed at location Y"), or the future change might dictate that design objectives should be revisited if a certain condition is found in the data (e.g., a decision rule that states "if results deviate more than 15% from expectations for COCs, then a new study is required to evaluate the conceptual model that was used to develop the LTM network"). These future decisions can use the optimal LTM network design tools presented here, as well as using these tools for the initial design.

In an optimal LTM network, when a sampling round occurs or when new results are available, those data should be analyzed to determine whether the objectives of the LTM system are being met and whether the LTM network operation or implementation can be changed or optimized based on the results. This analysis need not be comprehensive at the completion of each data collection, but could consist of, at a minimum, comparing measurement results with established trigger conditions. When a set of predetermined conditions occurs, various decisions could be made or actions taken. Questions such as "can sampling frequencies be reduced? can the number of analytes be reduced? are new sampling technologies available? do the results support the use of indicator samples? are examples of the kinds of issues that could be the focus of these predetermined actions. This approach allows for a continual evaluation of the system and provides a mechanism for the system to adapt to changing conditions. The tools presented in this report can aid in making a number of these decisions and will provide continual optimization of a LTM network.

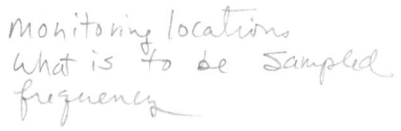

Monitoring locations
what is to be sampled
frequency

2.5.1. LTM Physical Design Issues. The physical design of LTM networks involves, at the most general level, the location of monitoring points, what is to be sampled, and how frequently these points should be sampled. The goal of an optimal design of a monitoring network is to reduce one or more of these parameters while still meeting the objectives of the LTM system. Optimizing a system to reduce these parameters is not a simple task. Sampling location decisions can be influenced by remedy selection, geologic heterogeneity, contaminant types, hydrogeologic conditions, and potential future needs. Sampling frequency can be influenced by sampling methods, sample cost and accuracy (screening or indicator samples), contaminant types, rates of groundwater flow, potential risks, and various other factors. The impacts of these issues can be modeled to determine the sensitivity of optimal configurations to them. Correctly establishing the objectives of an LTM network and the factors that can influence the optimal design of such a network can be difficult.

Many of the tools for the network optimal design require some sort of predictive model of subsurface groundwater flow and/or contaminant transport. These models might be very simple, such as regressions of head observations and one-dimensional analytic solutions, or full-blown time-varying three-dimensional numerical analysis codes. The case studies section provides some examples of how models can be used and integrated with optimal design tools. Although subsurface flow and transport models can be important for the optimal design of LTM networks, it is beyond the scope of this document to provide detailed information on the types or capabilities of the various models that are available.

2.5.2. Other Decision Parameters for LTM. Typically, physical design parameters, such as sampling well location and sampling frequency, are the focus in optimal LTM design; however, other decision parameters are involved. Beyond identifying these important needs in this section chapter, we have not addressed them in the remainder of this document because they are beyond the scope of the committee.

One particular area that is important for optimal design is the type of sample used. This topic is vast and rapidly changing as new sensors and field techniques are developed. New analytical techniques are also being developed. This is one area that may significantly change the way in which LTM systems are designed and operated in the future. Although new techniques and approaches are continually being developed, samples can be categorized into some general types:

- discrete samples recovered from an observation well
- sensors (the outputs of which may used to indicate the presence or absence of a contaminant or to quantify the concentration of contaminants present)
- indicator samples, which do not directly measure the contaminant of concern, but rather another parameter that is easier or more cost effective to measure (can be obtained by either direct or sensor methods)

13

- geophysics, remote sensing, and methods used to indicate the potential presence or concentration of a contaminant

These are general categories and, within each, there are a large number of specific technical topics and issues. A detailed discussion of how to use and implement them is beyond the scope of this report, but Chapter 6 discusses the need to address these emerging technical topics and issues in future research and technology transfer.

2.5.3. LTM Objectives. As noted previously, identifying the objectives of an LTM network is one of the most important steps in optimizing LTM designs. Without clearly stated objectives, the importance of different sampling locations and types cannot be identified using the methods presented in this document. The objectives can vary: regulatory, business, human health risk, ecological risk, and aesthetic objectives, as well as objectives that might require the optimal use of space, construction time, or operational cost. A design objective might be one that minimizes operation and maintenance costs with little regard to capital costs, or the reverse. All of these objectives are valid, and, for a given site, a number of competing objectives might need to be considered.

The objectives presented in this section of the chapter are physical objectives that can be addressed during LTM system design. These objectives should be included in the conceptual model of an environmental restoration site for design of LTM systems. We would like to stress that the objectives presented here are not the only objectives that can, or should, be considered in the design; however, the objectives presented can be explicitly included in the LTM methods presented in this report. These objectives will be used in Chapters 4 and 5 to aid in classifying methods and field applications.

Objective 1: Sample background. Federal regulations require that groundwater-monitoring wells be placed up-gradient of contaminant sources or in other locations to provide background information about local groundwater conditions. These wells are used to measure background levels of contaminants for comparison with down-gradient wells that may be affected by the source of contamination.

Objective 2: Detect release. Monitoring can be used to detect releases of contaminants from landfills or other sources before they adversely affect human health or the environment.

Objective 3: Verify flow containment. At sites with existing contamination, monitoring ensures that groundwater flows are contained, thus preventing migration to uncontaminated areas.

Objective 4: Verify plume shrinkage. When a remediation system is in place, monitoring is used to verify whether the existing plume of contamination is shrinking, which provides an assessment of the remediation system performance.

Objective 5: Verify plume stability. In some cases with recalcitrant or continuous low-level sources of contamination (such as sites with residual nonaqueous phase liquid contamination), the monitoring objective may be to verify that the contaminant plume is stable (i.e., not moving) and will not cause further contamination. This objective is similar to Objective 3 (and in fact, Objective 3 is often a surrogate for Objective 5), but in this case, the focus is on measuring contaminant concentrations, not just flows. This type of monitoring is particularly relevant at sites where natural attenuation is being used to achieve plume containment or remediation.

Objective 6: Monitor compliance after closure. After remediation has been completed and a site has been closed, a period of compliance monitoring may be required or desirable to ensure that contaminant concentrations remain below acceptable levels.

Objective 7: Verify or correct site models. As part of the remediation design process, analytical or numerical models of the site are often developed to predict the effectiveness of remediation options or to provide information for risk assessments. As data are collected during long-term monitoring, the data can be used to verify or, if needed, correct the model. This process helps confirm whether the predictions made in the remediation design are still valid.

Chapter 3: Environmental Data Management

A substantial loss of environmental restoration monies results from a lack of proper data management. How many sites are you aware of that suffer from the following short-comings: hard-copy data with little or no quality control information; the database cannot be used for deriving useful information; the data reside in a database that has been poorly managed? Under such circumstances, the data are too difficult to use and it is easier to resample and reanalyze. When the database is poorly managed, the data are often useless because important pieces of information may be missing or incorrect. There are many wonderful tools available to analyze and model data, but unfortunately up to 90% of the time spent will be used for cleaning up the data to make it useable for such tools. Patricia Ottesen, of the Environmental Restoration Division's Information System Management Group, offers the aphorism "the most expensive sample ever collected is the one whose results are never used." This does not mean that money can be thrown at data management and all of these problems simply go away; to manage data properly requires planning, adequate support, and a long-term commitment to a data management program.

The real importance of a data management system is to provide the end user with a consistent data set of known quality A good data management system should be one that is modeled according to how the data are collected and processed, is very well documented, has specifically defined data elements, and has supporting documentation. The latter attribute—supporting documentation—includes items such as quality control data that are carried along with the analytical data and metadata (information about the data).

Data sets tend to improve the more often they are used. As errors and inconsistencies are identified and corrected, the data set improves. A good system will achieve this because its design promotes use, and the ultimate value of data are, of course, in their use rather than their storage. The development and use of Internet access tools and existing environmental data management systems can help reduce the effort and cost associated with setting up a data management system.

This chapter is intended to provide a summary of some important data management topics. It will not go into great detail, but should provide readers with enough information to get started in managing their data. Common misconceptions about data management will be addressed, and readers will be provided with enough references to further investigate areas of specific interest.

3.1. Data Needs, Quality, and Assessment

When project data are already being managed in a database, staff members generally feel as if the data are being well managed. But how can that be verified? This section will try to address questions about existing and potential data management problems by discussing data needs, quality, and assessment.

3.1.1. Data Needs. When a project begins, two initial questions to be answered are as follows.

Are the correct samples being collected? The use of decision quality objectives (also known as *data quality objectives*; DQOs) (EPA, 1994) helps to ensure that the samples collected and analyzed will provide the data necessary to address the projects needs and that the data set will be of the appropriate quality. The DQO process is a seven-step method for defining quantitative and qualitative criteria for determining when, where, and how many samples (measurements) should be collected and with what desired level of confidence. Using DQOs can save more time and money than any other step in the entire environmental restoration process (Cave et al., 2001). To aid in the implementation of the DQO process, there is now a free software program called *Visual Sample Plan*. This program was developed by the U.S. Department of Energy and can be downloaded from the website http://dqo.pnl.gov/index.htm. Another site that is very helpful in finding information on the issues related to the DQO process is the Hanford website http://www.hanford.gov/dqo/related.html.

What types of data need to be collected and managed? Using the DQO process should help determine what data are necessary to manage. However, it is still important to ask all of the data users (chemists, engineers, hydrologists, geologists, biologists, project managers) what data they need and how frequently they use it. The ultimate goal is to manage only the data and quality control data that are needed. Associated quality control data should always be managed along with the original data; quality control data provide the information that allows a project to establish the quality of a data set.

3.1.2. Data Quality.
Data quality within a data management system (DMS) can be discussed from two very different points of view: the data manager's and the data user's. The manager focuses on how the data are processed in the DMS to ensure a quality data set, and the user is concerned with the quality (the precision and accuracy) of the data that are in the DMS (e.g., spatial, analytical, field data) and how the data are acquired. This section will discuss both aspects of data quality.

From a data manager's point of view, data quality is governed by the concepts of completeness, correctness, and suitability for use. Complete data sets and the user's knowledge about the current stage of data completeness are very important for proper interpretation of the data. Care must be taken to ensure that the data entered are correct; otherwise, queries and reports will yield results of questionable value. Even with correctly entered data, there may be occasions when the data are not suitable for use. For example, an analytical result for a sample may be correct in that everything done to produce and enter the data into the database was done properly. However, if associated quality control data indicate that some error may have been made in the analysis of the sample batch, the data point is unsuitable for use. Thus, a common way of handling this is to maintain this information in the database. When queries are performed, the ability to determine whether data should be used or not can be flagged for the user's attention. Many applications currently in use have the foundation in the underlying database. Bad-quality data may damage the application performance seriously (Garbajosa et al., 2000). The authors of this report recommend that a data quality process with a successful implementation have three parts: (1) data will be consistent with the real world, (2) transactional systems will be coherent, with a reliable behavior and capacity for historical projection, and (3) the historical files will contain data about the data (metadata) and context data.

One of the best ways of ensuring that the data will be complete, correct, and suitable is to correctly model the data management system in the beginning. A data model is a conceptual representation of the data structures required of a database. The components of this representation are objects or entity sets, attributes (the values describing some property of the entity), and relationships (how the entities are connected). The use of data modeling tools has significantly improved this process. Regardless of the method used to design the data system, there are several dimensions of quality to take into account. According to Reingruber and Gregory (1994), the five dimensions of quality are conceptual correctness, conceptual completeness, syntactic correctness, syntactic completeness, and enterprise awareness.

Conceptual correctness implies that the data model reflects the business concepts accurately. This is one of the most difficult aspects of assessing a model's overall quality, and it will have a great effect on the success of the system. Conceptual completeness means that the model contains objects adequate to describe the full scope of the business that the model is supposed to represent. Syntactic correctness implies that the objects contained in the data model do not violate any of the established syntax rules of the given language. Misrepresenting the relationships between tables in a database can lead to redundant data or to difficulty in querying. Syntactic completeness also refers to rules in place to ensure that modeling is done properly when following standards such as those of the National Institute of Standards and Technology.

From the data user's viewpoint, good-quality environmental data starts with a good-quality environmental program. This is a program that has quality assurance (QA)/quality control (QC) guidance documents and procedures in place. The U.S. EPA's QA/QC requirements for an environmental program are well documented in the EPA Quality Manual for Environmental Programs (EPA, 2000). This document provides requirements for the conduct of quality management practices, including QA and QC activities, for all environmental data collection and environmental technology programs performed by or for the EPA. This document can be downloaded via http://www.epa.gov/quality/qs-docs/5360.pdf. The main goal of having established standards and procedures is to create consistency. High-quality data are produced and managed by the consistent use of well-established standards and procedures.

This section cannot begin to touch upon all of the areas that affect data quality; however, the three areas that need the most attention are spatial data, analytical data, and field data. Good starting points for information on data-quality issues are the U.S. EPA (http://www.epa.gov) and the U.S. Geological Survey (http://www.usgs.gov) websites.

3.1.3. Spatial Data. Spatial data are affected by several factors, among them the use of consistent procedures and standards to ensure accuracy and precision, and a consistent effort to minimize sources of error.

Many helpful Internet sites provide spatial data information, standards, and workshops, such as the technology center for facilities, infrastructure, and environment at http://tsc.wes.army.mil/products/tssds-tsfms/tssds/html/sdsdocin.asp; the Geological Survey's site for the National Spatial Data Infrastructure, at http://mapping.usgs.gov/nsdi; and the Integrating Metadata Education Strategies site (information on data-quality measurement and assessment, error, accuracy, precision, and managing error), at http://www.sdvc.uwyo.edu/metadata/quality.html#gc. However, what if you already have graphical information system (GIS) data that you know contains topological errors. A method for correcting such errors has been described by Ubeda and Egenhofer (1997); this approach has three parts—the definition of errors, how to check the database, and how to correct the errors. The typical sources of error are as follows (Foote and Huebner, 1996).

- Age of the data: Data sources may be too old to be useful.
- Map scale: This is very important for determining data's fitness for use.
- Positional accuracy: A large number of factors contribute to positional accuracy, including the source materials used and digitizing methods.
- Density of observations or level/method of sampling: An insufficient number of observations may not provide the level of resolution (related to scale) required for a particular application.
- Sources of variation in data: Sometimes one data set is compiled by more than one person. A range of measurement errors can be introduced by faulty observations, biased observers, or the use of different equipment.
- Numerical errors: All GIS data are composed of coordinates, which are stored as numbers with a certain level of precision as allowed by the computer. Variation in the amount of significant digits handled by the computer or by the GIS software can cause rounding errors that may not be noticeable during processing but can become apparent when results are tabulated.
- Topological errors: Overlaying multiple layers of data in GIS can result in problems, such as "sliver polygons," or virtual data that may be difficult to detect from real data.
- Classification problems: Defining appropriate class intervals can be a subjective process that lends itself easily to bias.
- Digitizing and geocoding errors: These errors include poor source map material, faulty registration, physiological errors of the digitizer (sloppy digitizing), scanning errors, and means for correcting distortions in scanned materials or aerial/remotely sensed images.

3.1.4. Analytical Data. One of the most frequently questions asked in regards to analytical data is, how can I be sure I am getting good-quality analytical data from a laboratory (Taylor, 1987)? The following things can be done. Make sure the laboratory is certified, and obtain references; audit the laboratory (Garner et al., 1992; Bens et al., 1997); validate the data from the laboratory (EPA, 1985, 1988, 1991); participate in an interlaboratory comparison program (Farrar and Long, 1997); and use performance evaluation (PE) samples. The use of DQOs will also help in acquiring good-quality data by defining the specific project goals and the level of quality needed to support those goals. It is also helpful to understand the basics of analytical procedures (Cave et al., 2000; Clesceri et al., 1999; Anné, 1997) as well as the uncertainty and detection limits (Anné, 1992) associated with specific analysis. An interesting discussion on detection limits is given in the "Method Detection Limit: Fact or Fantasy" (Burrows and Hall, 1997).

Of all the checks that can be done, double-blind PE sampling is probably one of the most effective. Performance evaluation samples can be administered by two methods, blind or double-blind (EPA, 1997). When a PE sample is blind, the laboratory is aware that the sample is for PE purposes but does not know the chemical concentration levels. When a sample is double-blind, the PE sample is submitted as part of a field sample shipment. In this situation, the laboratory is not only unaware of the concentration levels, but it is also unaware that the sample is for PE purposes. The results of a double-blind PE sample are a direct reflection of how the laboratory is performing. Double-blind PE samples should be sent to a laboratory on periodic bases—quarterly if possible. The use of double-blind PE samples is expensive;however, it is frequently the only check that detects a problem.

Whenever possible, an electronic data deliverable is preferred over a printed report, which must be manually entered into the data management system. An example of an electronic data deliverable is shown in Appendix A, which consists of a list of elements, including results and the necessary QC data. Another good example of an electronic data deliverable is given in Arsenault and Legg (2001); this paper provides examples of good electronic data deliverables and does not promote the use of just one standard electronic deliverable. The number of elements required in an electronic deliverable depends on the needs and goals of the project.

The amount of error in the chemical analysis of compounds can be estimated; however, it requires effort and expense. Performance evaluation samples can provide an error range for the user of the data. The amount of error can depend on many factors, such as the analyte, the matrix, the analyst, and the sample handling (Clesceri, 1999). Another publication that provides an excellent presentation on the use of PE samples is "Lessons Learned from Performance Evaluation Studies" (Forman and Vitale, 1999). This paper discusses some of the issues that can arise with different matrixes and analytes. The report "Data Quality Objectives and Criteria for Basic Information, Acceptable Uncertainty, and Quality-Assurance and Quality-Control Documentation" (USGS, 1998) presents an overview of the difficulties associated with determining the level and sources of error. In most cases, the analytical error is small in comparison with the sampling error and environmental variability (USGS, 1998). The one major exception is with volatile organic compounds (VOCs) in a soil matrix (Jenkins et al., 1993); sample mishandling and length of the holding time can produce large errors.

3.1.5. Field Data and Sample Collection. The best way to produce the good-quality field data and to obtain the most consistent field samples is with the use of standard operating procedures performed by a well-trained staff. The use of duplicate samples taken in the field can provide information on sampling variability for water samples. Duplicate soil samples can also be taken; however, the soil's heterogeneity will tend to cause large variations in the analytical results. If there are more-basic questions to be answered, such as what sampling technique should be used and what analysis is most appropriate, then the EPA's Field Sampling and Analysis Technologies matrix (http://www.clu-in.org/pub1.htm) is very helpful in answering these types of questions.

Trying to estimate the amount of error due to the sampling process is a very difficult, and in some cases, nearly impossible task. It is so difficult because no one really knows the true value of an analyte before it is sampled. Estimating the amount of error in a water matrix (USGS, 2000) is an easier task than trying to the estimate the amount of error in a soil matrix. Soil matrices have many characteristics that complicate defining and quantifying the sources of error (e.g., soil heterogeneity, sorption, surface reactions, dissolution and precipitation reaction). The sampling error for nonvolatile analytes in a soil matrix, such as many types of metals, can be estimated by the use of field duplicates outlined in the report by van Ee et al. (1990).

Field sampling protocols should also minimize the potential for cross-contaminating the sampling locations. Dirty sampling equipment can contaminate soil or groundwater at a location that was previously clean. Such a contaminant can lead to an erroneous interpretation of the spatial extent of subsurface contamination, which in turn could substantially inflate site remediation costs. The potential for cross-contamination can be minimized by using disposable or dedicated sampling equipment at each sampling location or by thoroughly cleaning equipment when moving between locations. No cleaning agent is appropriate for all contaminants: the agent used at a particular site should be capable of removing contaminants previously encountered or anticipated at the site. Running purified water through cleaned sampling equipment and collecting and analyzing the sample can check the quality of an equipment-cleaning procedure. To further reduce the chance of cross-contamination at an established monitoring well network, sampling should start at (anticipated) cleaner locations and move to progressively more-contaminated locations.

3.1.6. Data Assessment. A data assessment allows the data user to determine whether there are enough data to make the necessary decisions and whether the data are of the appropriate quality. There are now several documents and software programs that can help in the data assessment process. One of the most comprehensive documents on data assessment is EPA's Guidance for Data Quality Assessment: Practical Methods for Data Analysis (EPA, 1998). This document covers a five-step process for assessing the quality of a project's data and describes many statistical tools used for data analysis. It provides a tremendous amount of information, step-by-step instructions, and many useful references.

3.2. The Qualities of a Good Data Management System

A good data management system (DMS) provides both a data loading tool and a data management and reporting system for environmental data associated with restoration and monitoring programs. Also available should be a tool for importing, translating, and exporting analytical data in various electronic data deliverable (EDD) formats, thus linking laboratory results to site information and enabling enhanced data reporting capabilities. The DMS should be able to manage analytical laboratory data with information related to sites, locations, soil borings, lithology, well installation and monitoring, soil geotechnical laboratory data, field sampling and field tests, chain-of-custody data, and regulatory requirements. An electronic project archive of known quality, with historical data that are easily accessible by different parties for use in future environmental projects, is a key element of a good system.

The choice of the appropriate underlying database is critical for a good data management system. Allowing for optimum flexibility, a data-driven system allows the data elements to be added or adjusted to meet a project's specific needs. Such potential changes to the system should encompass all major components of the software, including entry screens, import/export functions, security/revision tracking, and consistency-checking routines. This flexibility makes it possible to easily migrate or add information to an existing system. Hence, if an organization wishes to keep an existing system's design and content, the data from the original system could be moved into a data-driven system without translating, reorganizing, or renaming the original data.

Documentation is another important feature of a good DMS. A data dictionary contains a list of all files in the database and the tables, names, definitions, and types of each field. Data dictionaries do not contain any actual data from the database—only bookkeeping information for managing it. A data dictionary is a document that is also necessary for the translation of data. In the Installation Restoration Data Management Information System data example, which is described later in this chapter, the IRDMIS Data Dictionary was used to help in the translation. Although the IRDMIS documentation was complete in its description of the informational fields and coded values, its discussion of the data relations was incomplete in terms of describing what makes a record of information unique (primary key), as well as describing how the groups of data link together (data relations). What makes the data records unique and how the data groups are linked are essential to a user's understanding and interpretation of the data stored within the system. For example, if a geospatial location has multiple location names, how does one location name connect to another to pool data concerning a single geospatial point? Because the system's structure documentation is the basis for any database implementation, the lack of this documentation is a significant barrier to data interpretation.

Data-entry screens allow manual input of geotechnical, laboratory, field, groundwater, and lithological data, as well as project management and regulatory information. The entry screens provide a means of interactively viewing and modifying the DMS. These screens facilitate consistent data entry by checking values for logicality entry (e.g., the beginning depth is less than the ending depth). The system's modifiability and dynamic tasking accommodates user preferences, particularly organization-specific information and business practices.

An import function consists of tasks automated to locate and configure import files. Import validation procedures need to be provided to ensure that the imported information is structurally sound. Users need to be alerted to structural inconsistencies via some type of online error report generated by a consistency checker. A good DMS allows for the import of various electronic deliverable formats (EDFs) into the same DMS. The enABL Data Management System of Arsenault-Legg, Inc. allows for the import of the Electronic Data Format (EDF), Installation Restoration Data Management Information System (IRDMIS), Environmental Resources Program Information Management System (ERPIMS), Navy Electronic Data Transfer Standard (NEDTS), and the Environmental Management Electronic Data Deliverable (EMEDD) formats. Other EDFs are available, such as hierarchical or XML; however, for these formats to be practical there will be many issues that have to be addressed:

- Most analytical laboratories will not be able to directly produce a hierarchical or XML format. A laboratory typically has its data in a relational system to begin with, and using a hierarchical or XML format will require the data to be converted to a new format. The XML format as compared to an ASCII file is very hard to visually verify.
- The file sizes in the hierarchical or XML format are extremely large and very difficult to manage. File sizes could be >1 MB for just analyte from one sample.

An export function allows the generation of reports and export files based on user-defined criteria. Also, set-up functions utilize screens that allow the selection of restrictive criteria, database query routines, and destination selection (i.e., interactive browse, reports, and data exports).

Security/revision tracking functions allow for restricted-access accounts and the tracking of data modification. The DMS administrator has the ability to select user rights to such functions as read-only, data modification, report generation, and import functions. A good system tracks all revisions and is able to generate reports showing data modifications of imported and manually entered data, the user modifying the data, and the date and time that it was modified. It is important to be able to query and report on the revision tracking.

Consistency checking locates files, selects tables for review, and initiates electronic data checking. This procedure ensures that the data are structurally sound before being placed into the DMS. A user should be alerted to structural inconsistencies via an error report.

The elements in the previous sections are the foundation for a good DMS; however, it is the easy access to data that creates a useful system for the users. The Environmental Restoration Program of the Lawrence Livermore National Laboratory (LLNL) has been using Internet access tools with their DMS since 1995 (Canales and Ottesen, 1996). This system has allowed technical personnel, regulators, and the community to have controlled access to the LLNL's site data. The access is controlled by the use of accounts. Another DMS, the EDMS, has also added Internet access tools to their system.

Web tools and wizards assist with the management and dynamic reporting of environmental data that can be accessed by multiple approved users from anywhere in the world. These reports are accessed through the web tools via either Internet Explorer or Netscape. Reports are generated using the web tools that have been added to more fully utilize data contained within a DMS. Generally, minimal training in the use of these tools is required. Web tools are built to retrieve specific types of data, such as summary data reports and statistical analysis reports. These tools can also provide visualization of data, such as graphs and contour plots. The following are descriptions of just a few tools that are attached to the LLNL's DMS and to EDMS. LLNL presently has over 30 different web tools, and EDMS has over 15 tools. Once the initial system is in place, additional tools are easy to create. To see an example of these types of Internet access tools, a demonstration web site has been set up at http://www.arsenaultlegg.com/webtools. Tools available on the controlled LLNL web site include the following.

- GIS Contouring. Topographical map information can be imported to enable GIS plots where such mapping applications are available. The GIS contour plots can be generated using specified base maps and underlying data stored in the DMS. The generated plots can be zoom-enlarged for detailed viewing.
- Cost-Effective Sampling. This web tool enables cost-effective sampling (CES) (Ridley and MacQueen, 1995) analyses and reporting. The CES statistical tool utilizes data residing in the DMS to provide frequency recommendations for routine monitoring programs. The system allows real-time adjustment of constants and data ranges. Checking will occur to verify the statistical viability of data. Wizards are provided to assist in use of this tool and the creation of CES reports that may be generated from the data session.
- Time Series Graphs. This web tool enables time series data to be viewed in graph form. Users can enter a time interval, specifying start and end times, in order to easily see the changes of a parameter over time. The Time Series Graphs can display up to four customized data series in line graph format. These reports are generated from data residing in the DMS that can be viewed in table format or downloaded to a Microsoft Excel spreadsheet.
- Data Extraction/Reduction Reports. Data Reduction Reports can be generated to display specified data. The selected data can also be downloaded to an Excel spreadsheet for further manipulation.
- Project Management Reports. Wizards guide users in the creation of pertinent project management reports that determine site status, evaluate project schedules, and track regulatory requirements.

In the preceding discussions, Lawrence Livermore National Laboratory's DMS and the EDMS were referenced frequently as examples. Based on experience with these systems, both are well designed, well maintained, and able to handle large amounts of data. They have provided reliable service for many years. Information about EDMS is available at http://www.arsenaultlegg.com.

3.3. The Creation of a Good Data Management System

The creation of a new data management system is a great effort. Failure rates can be high if the appropriate people and tools are not involved in the process. The public tends to hear only about the very expensive failures, such as the California Department of Motor Vehicles system.. Data management needs to be a well-integrated part of any program to ensure that access to the data is successful.

The requirements for designing a good data management system can be broken down into the personnel required to accomplish this objective and the appropriate choice of a DMS capable of handling the data needs. Frequently, systems are created by a group of individuals with little understanding of the data that will be stored in the system. It is not enough to just know the elements that need to be stored in a system. To create a truly good system, there needs to be an overlap of understanding between the data users, who are presumably knowledgeable in the study area, and the computer support staff.

To create a good data management system, you need an individual(s) with a complete and thorough understanding of the data and how the data are processed. This is the person(s) who should be trained in data modeling and basic data management and who may also have relational database development experience. This person or persons should be responsible for designing the model for the system. If this individual is not experienced in database development, then it is also necessary to have an individual(s) who has an extensive background in database design with some knowledge and understanding of the data that will be stored in the system.

The other aspect to be considered is the choice of an appropriate DMS. It must be capable of handling the data requirements of the program. It also must be able to handle the appropriate data elements to cover the project's needs (e.g., the need to contain objects such as pictures). It should be built on a good entity relationship diagram in which the relationships are correctly modeled and be a flexible data-driven system. A data-driven system allows for additions and field changes without having to reprogram the entire system when it grows beyond original expectations. The software and platform necessary to build and successfully use the data management system must be identified. This can be either a desktop database management system (such as Access or dBase) running on a PC, or a larger SQL-compliant relational database management system.. Finally, it must be easy to use. Tools that allow the users easy access to the system are more likely to be used. The ability to access the database is not limited to only those few with database knowledge, but to a larger group of users.

3.4. Spreadsheet Versus Databases

There are many in the environmental business who believe data management can be achieved through the use of spreadsheets. Although spreadsheets provide excellent tools for analyzing data, they fall short of being able to properly manage data sets of any real size. The following is a short summary of why a database is a better tool for managing data (Patten, 1998).

- *Data sharing.* Although modern spreadsheets allow the sharing of workbooks over a network, the ability to control shared data sets is much easier in a database. In a spreadsheet, shared workbooks are made available to multiple users, but the control of individual columns of data is limited. Modifications made by multiple users to a cell in a shared Excel workbook must be accepted or rejected by the owner of the workbook. This makes it much more difficult to ensure a consistent data set. It is very easy in a database to achieve field-(column) level control of data.
- *Security.* Along with the ability to better share information is the ability to better secure it. You can protect private information better with a database. The database administrator can set up user rights down to the field level. This ensures that shared data sets can be viewed, modified, or updated only by individuals with the proper authority and, presumably, knowledge. This is not

true in a shared spreadsheet, where all networked users see the same version of the data.

- *Efficient.* Using relational databases, you link related tables to minimize duplication. For example: In a well sampling program, there is a set of chemical results for a particular set of wells. These wells have been sampled and analyzed for all of the same analytes. In a spreadsheet, you would have to duplicate the information for each well and each analyte (method ID, date, laboratory). This is referred to as data redundancy. A properly modeled relational database will have been normalized to reduce this redundancy, and this information would be entered just once.
- *Better reporting.* It is easier to create detailed reports in a database. Modern database tools have allowed the user to pick only the fields desired, along with the grouping and sort order, and a report is instantly created.
- *Greater capacity.* Databases have the capacity to hold more records than spreadsheets. Spreadsheets are limited in the number of rows and columns. Large database tables can generally handle an unlimited number of records, although the ability to query these tables can result in slow performance if allowed to grow too large. A plan for archiving historical data when a table reaches this point should be part of a good data management system.
- *Data integrity.* The ability to place constraints—better controls that restrict data to specific ranges and default values—is easier in a database. In a spreadsheet, formatting options for cells control entries to a certain extent. Simple logic statements can also be added to achieve some entry checking. In the case of using logic in cell formulas, many users do not have the level of sophistication required to create them. Putting constraints on fields in a database can be achieved through the use of data modeling software, through application programming for entry screens, and in the original field definitions to some extent. This makes databases a superior choice for handling large data sets.
- *Easier to maintain.* What makes spreadsheets so simple to use—the fact that the programming and the formatting are together on one page with the data— also makes them hard to maintain. When spreadsheets are used for database activities, such as list management and process tracking, users often spend time fixing spreadsheets because the formatting has inadvertently been changed while making changes to the data. That problem does not happen with databases, where controls have been properly placed on data as it is entered.

Databases require a greater investment in training, but the return is greater than from any spreadsheet. But two circumstances provide good examples of when using a spreadsheet makes sense. When doing a one-time analysis, a spreadsheet is a fast tool for building a simple snapshot that might not be used again. The time spent modeling and creating a database would not justify the gain. When compiling information from diverse sources, a spreadsheet is a good choice. For example, if you want to compare data from your restoration site with another site that has similar contaminants, it may be a simple matter to import their data into a preexisting spreadsheet. It can be much more complicated to import into a database.

Dishman (1998) has come up with the following questions to help answer whether or not a project needs a database. If you answer yes to two or more of the following questions, then it may be time to think about moving data from a spreadsheet to a database.

- Do changes made in one spreadsheet force you to make changes in another? Although this can be accomplished by linking worksheets in the newer spreadsheets, many users lack this level of knowledge and will maintain numerous "standalone" worksheets. This requires them to keep all related information updated on different worksheets.
- Do you have several spreadsheets containing similar information? If data redundancy among spreadsheets requires you to keep updating several spreadsheets, it makes sense to think about keeping all of the information in one place, where the data are only entered once.
- Do you want some data to be hidden from some users? Although spreadsheets allow you to hide rows and columns, the ability to hide individual cell entries is difficult. In a database, field level control is possible. It is possible to control what is available to a user on a case-by-case basis.
- Can you see all pertinent data on one screen or is it necessary to keep scrolling? This may be an irritation only when working with a large spreadsheet. The ability to filter and sort has made spreadsheets easier to manage. Querying results from a database allows a user to easily get just the data they are looking for in one step.
- Are several people accessing the data at the same time? As discussed in the section on data sharing above, networked databases are intended for sharing. Record and file locking control updates and modifications, and ensure that data are not overwritten unintentionally.
- Do you have difficulty viewing specific data? Databases are intended to produce finely tuned data sets by the use of querying tools or structured query language.

3.5. Desktop Database Versus Server Database

With some idea of the qualities of a good data management system and how to create one, it is then necessary to decide whether the system will be a standalone or a server-based.. This decision is based on several factors: how much data needs to be managed, how many users need to access the data, and what type of performance is necessary. A desktop database (e.g., Microsoft Access, Lotus Approach, Corel Paradox, Borland dBase) can run on a single PC or share data among several users. A standalone system is typically easy enough for a novice to start building immediately and robust enough to build a multiple-user networked application. Larger enterprise database applications (Oracle products, Microsoft SQL Server, IBM dB2) store all data on a networked server. The applications that access the data can reside on the workstations or on the server. They can handle massive amounts of data and include features that make constant multiple-user access fast and reliable (Dishman, 1998). The choice of a database system often depends on economics. Enterprise database systems are much more expensive, and often require higher end computer hardware to serve the database. Newer web-based applications for databases are generally optimized for larger enterprise database systems and may narrow the choice depending on project needs.

3.6. GIS ≠ Data Management

"For 20 years GIS vendors have defined their products and their industry in terms of geographic analysis, cartographic visualization and graphic interface technology. . . . Database management has for most vendors been a secondary consideration treated with breathtaking ignorance and, occasionally, disdain—a paradox perhaps considering the trouble and expense incurred by clients in order to capture spatial information" (Seaborn, 1995).

A geographic information system (GIS) is a computer system capable of assembling, storing, manipulating, and displaying geographically referenced information, i.e., data that are identified according to their location. Practitioners regard a complete GIS as encompassing the operating personnel and the data (USGS, 2000). It is inaccurate for users to believe that if their data are stored in a GIS then the data are being properly managed. A better way to view a GIS is to consider it as a powerful tool for helping to understand and make decisions about the data stored in a data management system. As discussed in the sections on the qualities and the creation of a good DMS, up-front work is necessary when setting up a DMS and additional work is required for its maintenance. Frequently, the work required to set-up the DMS to accommodate a GIS is not handled appropriately owing to a lack of database training on the part of the individuals operating the GIS or simply owing to a lack of time. There are many helpful websites that provide information on GIS database concerns. A thesis on GIS database design and organization (Ismail, 2000) can be found at http://www.hbp.usm.my/Thesis/HeritageGIS/master\research\ Database.htm. Salford University has a lot of information concerning GIS database issues and database design at http://www.els.salford.ac.uk/geog/staff/nmt/fungis/hsd/ hsd5.htm.

The main purpose of a GIS is to display data—not manage it—and the model results (displays) will be only as good as the entered data.

3.7. Data Cleanup and Translation

Translation and cleanup of a data set must be done before entering into a DMS. If you are just starting a project, then the data collection efforts can be put into place at the beginning and translation may not be necessary. However, if you have hard copy or electronic data that needs to be entered into a DMS, then this section maybe very helpful. The following discussion is an example of the translation and cleanup of the Riverbank Army Ammunition Plant (RBAAP) site data from the Army's Installation Restoration Data Management Information System (IRDMIS) to the Environmental Data Management System (EDMS). Several difficulties related to the IRDMIS data set were encountered.

In general, the difficulties can vary in terms of the actual impact on the translating effort. To translate data from one system to another, the structures of the systems need to be known. In the case of the IRDMIS data translastion, no information was available on the database structure. The missing information included both the data relations (how one group of data links to another) and primary key identification (what makes a record of information unique within a data group). Without this information, any interpretation of the data within the IRDMIS system was, at best, an educated guess, and that was exactly how the IRDMIS data had to be handled.

Among the most frequently encountered data translation problems are the following.

- Informational fields documented as "required" by the data system do not contain data. (Some of these informational fields are basic to data interpretation such as matrix, elevation, and units of measurement).
- Some of the coded values within the data (which are used to represent information such as analytical method) are not documented in the data dictionary.
- Entries that are supposed to be limited to specific data tables, as dictated by the documentation, appear in tables in which they do not belong.
- Locations are identified by more than one name. Frequently, the connections between the data groups are not documented. If a location is named one thing in one data group and something different in the next data group, the pieces of information may not link via location, as would be standard in a geospatial system. This significantly hinders the interpretation of the data.

There is a key distinction that needs to be made concerning data translation and data "scrubbing," or "grooming." The data translation is the process by which the data are moved from one DMS to another, such as moving data from IRDMIS to EDMS. Data scrubbing/grooming is the process by which the data are examined and cleaned to make the data set usable. As in the example of the initial Riverbank data translation and data scrubbing/grooming, both were very labor-intensive processes. Subsequent translations of IRDMIS data will be much easier due to the automated translation tool that was developed during the Riverbank data translation. However, the data scrubbing/grooming will be just as labor intensive each time data from a new site are examined and prepared. Unfortunately, data systems that act solely as archives with little or no retrieval activity tend to contain data with internal inconsistencies and perplexing ambiguities.

3.8. Conclusion

Proper data management is important to the success of any environmental cleanup project. The value of high-quality data for making informed decisions is critical. Integrating proper data management throughout all phases of program development results in a system highly valued for its completeness and accuracy.

Providing customized tools and applications for accessing the data affords easier access to a large group of users and further increases the usefulness of the system.

Appendix: List of Elements in Environmental Management Electronic Data Deliverable for Analytical Laboratory Results
(Available online at http://emnamp.inel.gov/edd/emedd.html)

Amount Added	Counting Error Type	Lab Reporting Batch	QC Precision Limit High
Analysis Duration	Custody ID	Lab Sample ID	QC Precision Limit Type
Analysis Duration Units	Date Submitted	Matrix ID	Quantitation Limit
Analysis Type	Detection Limit	Original Lab Analysis ID	Quantitation Limit Type
Analyte ID	Detection Limit Type	Original Lab Sample ID	Reporting Limit
Analyte Name	Dilution Factor	Pay Item	Reporting Limit Type
Analyte Type	EDD ID	Percent Moisture	Result
Analyzed	EDD Version	Percent Recovery	Result Basis
Batch Date	Filtered	Percent Recovery Limit High	Result Derivation
Batch ID	Instrument ID	Percent Recovery Limit Low	Result Text
Batch Procedure ID	Lab Analysis ID	Percent Recovery Limit Type	Result Type
Batch Procedure Name	Lab ID	Percent Solids	Result Units
Batch Type	Lab Matrix ID	pH	Retention Time
Client Method ID	Lab Name	Preservative	Sample Type
Client Sample ID	Lab Procedure ID	Primary Result	Temperature
Collected	Lab Procedure Name	QC Linkage	Uncertainty
Comment	Lab Qualifiers	QC Precision	Uncertainty Type
Counting Error	Lab Receipt	QC Precision Type	

Chapter 4: Methods for Optimizing Long-Term Monitoring Designs

Optimal long-term groundwater monitoring design can be accomplished using a variety of approaches. The pertinence and relative merits of the general approaches and techniques depend on several factors, among which one can cite: (1) the scale of the monitoring program (local, intermediate, or regional), (2) the objective of the monitoring program (ambient, detection, compliance, definitive-action, or remedial monitoring), (3) the type of data (subsurface stratigraphy, water levels, and groundwater chemistry), (4) the nature of the contaminant process (for example, transport and fate of chemicals in the vadose and saturated zones), (5) the steady-state vs. transient nature of groundwater quality properties, and (6) the changing goals of a long-term monitoring program. Institutional, legal, and other site-specific considerations may also bear on long-term monitoring requirements.

Several approaches have been developed for identifying suitable monitoring plans. Most of the early work in monitoring design focused on methods for siting new monitoring wells. The problem of siting new wells for site and plume characterization or for plume detection at landfills and hazardous waste sites has been studied extensively, and a variety of methods is available. Two ASCE committee reports (ASCE, 1990a, b) and Loáiciga et al. (1992) provide reviews of many of these methods. Recently, methods have been developed for identifying sampling plans that minimize spatial and temporal redundancy in existing monitoring networks. Both sets of methods—those for siting new wells and those for identifying or refining sampling plans in existing networks—are covered in this chapter. In addition, this chapter contains a brief review of procedures for analyzing groundwater monitoring data (Section 4.2). Those procedures are an integral part of long-term groundwater monitoring and network design/alteration.

The methods summarized in this chapter are organized according to the types of tools, the level of complexity, and the objective of the long-term monitoring. A primary consideration in selecting an appropriate method for optimizing long-term monitoring design is the quality and quantity of pertinent data. For a site with a point-source of contaminants and a few (say, <10) monitoring wells, only the simplest of methods would usually be warranted. For a site with thousands of monitoring wells, utilizing calibrated three-dimensional numerical models and collecting large volumes of data each year, the most sophisticated optimization analyses would be feasible and most likely worthwhile. Table 4.1 summarizes the methods presented in this chapter by the level of complexity involved and data required.

Another major consideration in selecting an optimization method is the objective of the monitoring program. Most monitoring programs have multiple objectives and, within a network of monitoring wells, different wells may have different purposes. For example, monitoring wells that are located down-gradient of a contaminated area are typically sentry wells that are monitored to ensure that the contamination is not migrating into uncontaminated areas. This situation exemplifies detection monitoring. In remedial monitoring or in definitive-action monitoring, wells that are located within contaminated areas undergoing remediation are typically sampled to assess the progress of ongoing remedial activities and determine whether adjustments are needed. Table 4.2 provides a list of some typical LTM objectives (see

TABLE 4.1. Long-term monitoring optimization methods by level of complexity

Method	Few	Intermediate	Many
4.1. Rule-Based Methods			
4.1.1. Hydrogeologic method	X		
4.1.2. Rule-based trend methods (examples: CES; MAROS)	X	X	
4.2. Statistical Methods			
4.2.1. Statistical comparisons	X	X	X
4.2.2. Trend detection	X	X	
4.2.3. Geostatistical methods		X	X
4.2.4. Hydrogeologic–geostatistical method		X	X
4.3. Probabilistic Methods			
4.3.1. Kalman filter methods			X
4.3.2. Probabilistic simulation methods		X	X
4.3.3. Hierarchical method		X	
4.4. Mathematical Optimization Methods		X	X

Amount of Data and Information spans the Few / Intermediate / Many columns.

Section 2.5.3 for explanations of these objectives) and notes which methods may be most appropriate for each objective. This list is not exhaustive and other methods may be equally appropriate to achieve objectives at some sites. A summary of each method can be found in the following sections.

This chapter provides a conceptual map of why to choose a particular method for long-term groundwater monitoring design, plus introductory material on how to implement those methods in practical settings. For detailed algorithmic and computational formulae and software, the reader must consult the references cited at the end of this chapter.

TABLE 4.2. Optimization methods by monitoring objective commonly found in practice

Method	Monitoring Objectives						
	1. Sample Background	2. Detect Release	3. Verify flow containment	4. Verify plume shrinkage	5. Verify plume stability	6. Monitor Compliance after closure	7. Verify or correct site models
4.1 Rule-Based Methods							
4.1.1 Hydrogeologic method	X	X	X			X	
4.1.2 Rule-based trend methods (CES; MAROS)			X	X	X		
4.2 Statistically-Based Methods							
4.2.1 Statistical comparisons	X	X	X		X	X	
4.2.2 Trend detection			X	X	X		
4.2.3 Geostatistical methods	X	X	X	X	X	X	
4.2.4 Hydrogeologic-geostatistical method	X	X	X	X	X	X	X
4.3 Probabilistic Methods							
4.3.1 Kalman filter methods		X	X	X	X	X	
4.3.2 Probabilistic simulation methods		X				X	X
4.3.3 Hierarchical method		X				X	
4.4 Mathematical Optimization Methods		X	X	X	X	X	X

4.1 Rule-Based Approaches for Identifying Long-Term Monitoring Plans

4.1.1. Hydrogeologic Method. This approach is commonly used for long-term groundwater monitoring programs, especially in the detection and compliance monitoring of sites regulated under the U.S. Resource Conservation and Recovery ACT (RCRA). It relies on qualitative and quantitative hydrogeologic information. Let us clarify that the discipline of hydrogeology encompasses all approaches that deal with groundwater, including statistical ones. Thus, all the methodologies cited in this chapter could be classified as hydrogeologic. Nevertheless, to be consistent with previous work on subsurface monitoring (Loáiciga et al., 1992), the term *hydrogeologic approach* will be used to describe the case where the long-term monitoring program is determined based on the calculations and judgment of the hydrogeologist without resort to advanced statistical and/or probabilistic techniques. Specifically, the sampling sites (wells) and times are determined by the hydrogeologic conditions near the source of contamination.

For example, federal guidance documents published by the U.S. Environmental Protection Agency (USEPA, 1986) require, at a minimum, three groundwater monitoring wells located down-gradient and one well located up-gradient from a contaminant source. The documents further specify that up-gradient wells must (1) be located beyond the up-gradient extent of potential contamination from a waste site to provide groundwater samples representative of background water quality, (2) be located close enough to the waste site and screened through the same stratigraphic horizons as the down-gradient wells to ensure comparability of water-quality data, and (3) be of sufficient number to account for heterogeneity in background-water quality. Down-gradient monitoring wells must consider: (1) the distance to the contaminant source and the direction of groundwater flow. (2) the likelihood of intercepting potential contaminant plumes. and (3) the physical and chemical characteristics of the contaminant source and geologic conditions that govern the movement and distribution of contaminants in the aquifer. In practice, the sampling frequency varies from daily to annual depending on the perceived exposure hazard associated with a contaminant source. The exposure hazard takes into account the toxicity of contaminants, the velocity of groundwater migration and contaminant dispersal, and the proximity of source aquifers and sensitive environments. The sampling frequency may change over time as a contaminant threat evolves through containment, remediation, treatment, and closure phases.

The hydrogeologic approach is better suited for site-specific studies in which there is a well-delimited (potential) source of contamination. The geologic characteristics, shape, and orientation of the contaminant source, and local and regional groundwater flow patterns at the site determine the horizontal and vertical distribution of sampling locations. In particular, aquifer layering and fractures control the vertical placement of sampling points, which may consist of multiscreened wells or well clusters. Often, the hydrogeologic approach is used to detect pollution as soon as it migrates outside the confinements of a waste facility.

The hydrogeologic approach may also rely on analytical or numerical simulations of groundwater flow and contaminant transport for estimating plume geometry, defining probable impact zones, and carrying out *a priori* tests of the detection effectiveness of groundwater sampling networks (Buller et al., 1984; Hudak and Loáiciga, 1993; Hudak et al., 1995). Flow and transport simulations can also be used to establish appropriate setbacks for up-gradient groundwater monitoring wells (Hudak, 1999). A typical set of coupled numerical models that are widely used for this purpose are the flow model MODFLOW (McDonald and Harbaugh, 1988) and the transport model MT3D (Zheng, 1990). For more information on flow and transport simulation, see a groundwater modeling reference such as Anderson and Woessner (1992).

Figure 4.1 illustrates some of the principles applied in long-term monitoring according to the hydrogeologic approach. The aquifer in Figure 4.1 is a glacial deposit with two sand layers in hydraulic connection via a low-conductivity glacial-till aquitard. Bedrock is a granite aquifuge. The upper sand layer has a greater Darcian velocity in a southerly direction. The lower sand layer has a lower hydraulic conductivity, and its ground water is part of a deeper regional flow system with an easterly direction as shown by the arrows in the figure. There are two waste lagoons (W1 and W2) in the top sand layer whose bottoms are above the water table. Further hydrogeologic information includes the physical characteristics of the various formations, ambient water quality, and hydraulic-head contour levels for the upper and lower sand aquifers.

A tentative set of long-term monitoring locations could be as follows. Consider first the down-gradient monitoring wells 5 through 13 in Figure 4.1. The upper sand layer is tapped by five monitoring wells (5, 6, 7, 8, and 9 in the figure). Of those, wells 7, 8, and 9 are well clusters that penetrate the various formations (upper sand, till, and lower sand). Any leachate from the lagoons must pass the upper sand layer and till before reaching the lower sand layer. Therefore, fewer single wells (10, 11, 12, and 13, in the lower sand layer) are located along the eastern perimeter. These wells are screened in the lower sand layer.

The up-gradient monitoring wells are located along the northern (wells 1 and 2) and western (3 and 4) perimeters and are screened in the two sand layers to provide background water quality data to be compared with the down-gradient wells. The larger velocity and exposure of the upper sand call for more frequent monitoring in that layer, especially after heavy precipitation events. The up-gradient wells would likely be sampled at least quarterly to provide long-term insight on the seasonal variability in water quality. Loáiciga et al. (1992) provide guidelines for temporal groundwater quality sampling that consider the typical velocities of ground water in various types of geologic environments.

Quinlan (1990) outlined principles of hydrogeologically based monitoring in karst aquifers. The strong hydraulic connection between rainfall and cave recharge requires sampling frequencies that depend on the temporal characteristics exhibited in groundwater flow hydrographs throughout the aquifer. In this instance, the geometry of a cave network dictates the placement of monitoring wells.

The previous example demonstrates that the hydrogeologic approach to long-term monitoring is data intensive and well-suited for site-specific studies where the overriding goal is early detection of contaminants migrating beyond site boundaries. In addition to considering site-specific hydrogeologic conditions, detection networks should be capable of detecting contaminant releases originating from anywhere within the footprint of a landfill. Common mistakes made in practice include using too few wells and placing them too close to the boundary of a waste site. Typically, contaminant plumes are narrower near the boundary of a waste site, and therefore more difficult to detect, than they are further down-gradient of the waste site. On the other hand, detection wells should not be located arbitrarily far away from a landfill because a large part of the aquifer will have been contaminated prior to detection. Once established, water quality observations from a down-gradient groundwater monitoring network can be compared with those from an up-gradient network, within a statistical context, in order to determine whether the aquifer has been impacted (Gilbert, 1987; Gibbons, 1994; ASTM, 1996). If impacted, the monitoring objective would shift from detection to compliance, and additional wells may be needed to spatially characterize the groundwater contamination.

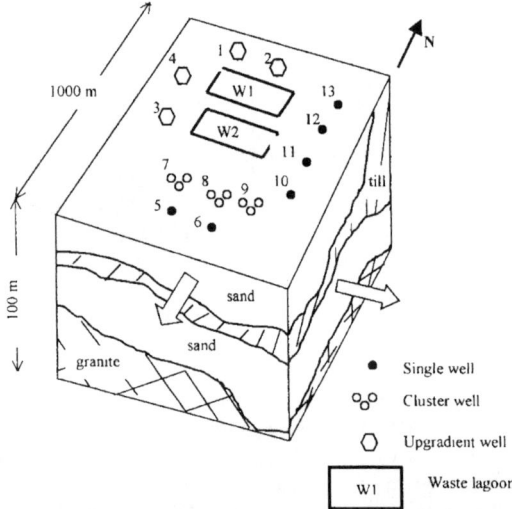

Figure 4.1. Monitoring according to the hydrogeologic approach.

4.1.2. Rule-Based Trend Methods. Another class of monitoring method combines professional judgment and experience with statistical methods into structured decision tree (or decision logic) frameworks for the purpose of developing long-term sampling schedules at existing wells. These rule-based trend methods follow a set of decision-logic steps (decision tree) that lead to specific decisions as to where and when to monitor groundwater. A decision tree is a convenient way of organizing a staged or sequential set of decisions of problems that involve uncertainty, a number of decision points, and alternative scenarios. An example of such a method is the cost-effective sampling (CES) technique (Ridley and MacQueen, 1995; Johnson et al., 1996). The CES technique produces a minimum-frequency sampling schedule for a given groundwater monitoring location that provides the needed information for regulatory and remedial decision-making. The determination of sampling frequency for a given location is based on trend, variability, and magnitude statistics describing the contaminants at that location. The underlying principle is that a location's schedule should be primarily determined by the rate of change in concentrations that have been observed there in the recent past. The higher the rate of change, whether upward or downward, the greater the need for frequent sampling. Conversely, when little change is observed, a relaxed schedule can be followed. The CES technique was implemented at Lawrence Livermore National Laboratory in 1992 and approved by the USEPA Region IX and the local regulators. Applying the CES methodology has reportedly produced a 40% reduction in the required annual number of routine groundwater samples at Lawrence Livermore, with an estimated savings of $390,000 per year (in 1992 dollars).

A second example of a rule-based trend method is the monitoring and remediation optimization system (MAROS, Aziz et al., 2000). MAROS is a rule-based method for developing sampling plans in existing networks. It uses the lines-of-evidence and weight-of-evidence approaches that are widely used in risk assessment (see USEPA [1996] or U.S. Navy [2001] for descriptions in the context of ecological risk assessments). MAROS is a public-domain software developed by Groundwater Services, Inc., for the Air Force Center of Environmental Excellence (AFCEE) to assist in the implementation of the AFCEE Long-Term Monitoring Optimization Guide (AFCEE, 1997). MAROS identifies a site-specific monitoring program for sites with monitoring objectives that involve tracking contaminant migration in groundwater. It uses the Mann-Kendall test and linear regression analysis to perform trend analysis in individual wells and uses these as lines of evidence for determining plume stability. It also identifies temporal sampling plans using a variant of the CES method and spatial sampling plans using a Delaunay-interpolation method. The latter method uses Delaunay triangulation and estimates of interpolation errors to identify the importance of existing monitoring wells in estimating the average concentration of a plume. For each well with observation data, the contaminant concentration is estimated using all of its natural-neighbor wells defined by Delaunay triangulation. A factor, which is defined as the standardized difference between the measured concentration at the well and the concentration estimated using its natural-neighbor wells, is then used to determine which wells provide little information about groundwater contamination.

4.2 Statistical Methods

4.2.1. Statistical Comparisons. These methods are used to make statistical inferences concerning (1) background well to compliance well comparisons, (2) comparison of compliance well data with a water-quality standard (e.g., a maximum contaminant level [MCL]), and (3) intra-well comparisons. These types of comparisons are common in—but not limited to—detection, compliance, and

37

remedial (or corrective action) monitoring (see Davis, 1994; Davis and Nichols, 1994a, b). Figure 4.2 illustrates a flow chart to aid in the selection of a statistical comparisons method when high concentrations of pollutants are caused by human activities and are otherwise found at trace or zero levels naturally in groundwater. Such as is the case at many RCRA facilities (see USEPA, 1989, 1992).

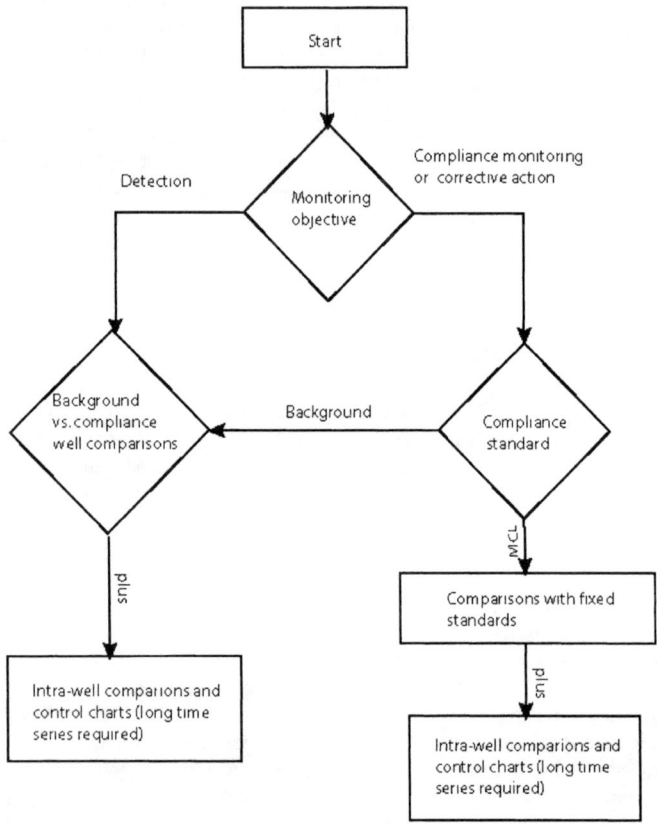

Figure 4.2. Flow chart for selecting a statistical comparison method (the MCL, maximum contaminant level, can be replaced by some other alternative contaminant level, ACL).

In Figure 4.2, one starts at the top and arrives at the first decision node, where, if the monitoring objective is contaminant detection at a specified location, then a background vs. compliance well comparison is applied. If, however, the monitoring objective is, for instance, corrective action, then one proceeds to the decision node in the right side of the chart. In the event that a compliance standard is specified as a maximum contaminant level (MCL) or alternative contaminant level (ACL), a comparison with a fixed standard is called for. Otherwise, the compliance standard is a background indicator and the chart branches to background vs. well comparisons. Both the left and right branches of the chart end with a possible intra-well comparison and the construction of control charts to test for temporal patterns.

4.2.1.1. Background well to compliance well comparisons. Data from detection, compliance, and remedial action wells must, in a variety of situations, be compared to so-called background well data. A background well is representative of natural groundwater quality characteristics, meaning not influenced by human activity. The background well to compliance well comparison methods are classified into the following (see USEPA, 1989): (1) analysis of variance (ANOVA) procedures and (2) tolerance and prediction interval comparisons.

(1) ANOVA procedures. These are used to compare several populations, typically in terms of their means. One-way ANOVA is applied when one factor (e.g., a contaminant source) may be responsible for interpopulation differences, whereas two-way ANOVA is appropriate when two factors (e.g., a contaminant source and the type of geologic formation) account for those differences. ANOVA procedures may be grouped into parametric and nonparametric procedures. The former assume that well data or that transformed data (generally concentrations of chemicals in ground water) are normally distributed, whereas the latter dispense with the normality assumption about the raw or transformed concentration data. Concentration data in groundwater are usually rendered approximately normally distributed when log-transformed. In any case, there are statistical tests of normality that can be used to establish or reject normality of a data set. ANOVA procedures can be used to test the null hypothesis that the concentration averages in nonbackground wells (detection, compliance, remedial action wells) equal the mean concentration in background wells. If the data from both sets of wells—background and nonbackground—are collected during the same time interval, then, one implements a one-way ANOVA test (either parametric or nonparametric). When the data in the background wells are collected during a period different from that that of the nonbackground data, then one must resort to two-way ANOVA tests. The latter are tantamount to an adjustment for seasonality differences between the well data sets. See details on ANOVA procedures and examples in Hollander and Wolfe (1973), Johnson and Leone (1977), Miller (1981), and USEPA (1989, 1992).

(2) Tolerance and prediction intervals. A tolerance interval is constructed from the data in background wells. The data from nonbackground wells are compared with the tolerance interval. If they do not fall within the tolerance interval, one concludes that there is contamination in the nonbackground wells. Tolerance intervals are most appropriate for use at contaminated sites with high degree of geologic homogeneity. They also require that the raw or log-transformed data be normally distributed. When the number of background well data (n) is equal to or larger than 20, a one-side upper tolerance interval (T.I.) is approximated by

$$T.I. = \overline{X} + z_{\alpha} S$$

where $\overline{X}$ and S are the mean and standard deviation of the background well data, respectively, and z_{α} is a standardized (i.e., zero mean, unit variance) normal variable such that $P(Z \geq z_{\alpha}) = \alpha$. The parameter α is called the *significance level* and is usually set equal to either 0.05 or 0.01. When the number of background well data is less than 20, the z_{α}-statistic is replaced by the $t_{\alpha, n-1}$, the so-called t-statistic with $n - 1$ degrees of freedom. If any observation from a nonbackground well exceeds the tolerance interval, it is concluded that the well is contaminated. If the logarithms of the background well data are used to construct the T.I., one must also take the logarithm of the nonbackground well data prior to comparing them with the T.I. Alternatively, the antilogarithm of the T.I. is taken and then compared with the nonbackground well data.

Prediction intervals (P.I.) are constructed so that the mean of each of M nonbackground well samples (each sample contains m concentration measurements) can be tested for evidence of contamination by comparing it with the P.I. Each of the M nonbackground samples is taken at a nonbackground well in a future time period; thus, there are M different sampling periods (usually in the future, as new data are collected for the test) at the nonbackground well. The P.I. is constructed based on the mean and standard deviation of n background-well data points. Each of the M nonbackground well means are calculated and compared with the P.I.: if a mean falls outside the P.I., this is evidence of well contamination. See details in USEPA (1989).

4.2.1.2. Comparisons with fixed standards. Comparisons with fixed standards are used to test whether groundwater concentrations are below or above a maximum concentration limit (MCL) or an alternative concentration level (ACL). The procedures for carrying out these comparisons can be classified as parametric or nonparametric. The former assume that the raw or log-transformed data are normal. The latter do not require normality of the concentration data. We describe herein a type of parametric confidence interval procedure to compare well data with fixed standards. There are, in addition, nonparametric confidence interval procedures. Another category of comparisons with fixed standards relies on tolerance intervals, which will not be reviewed here.

A parametric confidence interval (C.I.) is formed by first calculating the concentration mean $\overline{X}$ and standard deviation S from a nonbackground well (from m data points), and then calculating the C.I. as follows:

$$\left\{ \overline{X} - \frac{S}{\sqrt{m}} t_{\alpha, m-1}; \ \overline{X} + \frac{S}{\sqrt{m}} t_{\alpha, m-1} \right\} \tag{4.1}$$

If the lower limit of the C.I. exceeds the MCL or ACL for the monitored chemical in the nonbackground well, this is evidence that the well is noncompliant. If the comparison is implemented with the logarithm of well data, then one must take the logarithm of the MCL or ACL prior to carrying out the comparison with the C.I.

Alternatively, the lower and upper limits of the C.I. must be inverted by taking their antilogarithms and then compared with the MCL or ACL.

4.2.1.3. Intra-well comparisons. In this case, concentrations of selected constituents are monitored over time in a single uncontaminated well. This is effectively done with a Shewhart-CUSUM control chart for each well and constituent. The data for a constituent are plotted against time and inspected for trends and changes in concentration. The control chart method is recommended for uncontaminated wells only, with at least eight independent concentration measurements during a one-year period. Natural seasonal variations of a constituent's concentration must be removed prior to plotting the time series data in a control chart. By removing those seasonal components, when they exist, it is possible to assess temporal changes due to nonnatural causes, those that are of interest in the context of compliance monitoring, for example. Appendix A presents a simple procedure for removing seasonal components in concentration data.

The concentration data used in a Shewhart-CUSUM control chart are collected at different times and must be mutually independent. They, or a suitable transformed variant (e.g., their logarithm), are assumed normally distributed as well. Suppose that n independent samples of a constituent are collected at a monitoring well. The samples are collected at different times, mutually independent, adjusted for seasonal components, and normally distributed (after suitable transformation if necessary). The sample collected at time t_j, $j = 1, 2, \ldots, n$ ($n > 8$), contains n_j measurements, whose mean and standardized mean are $\overline{X}_j$ and $Z_j = (\overline{X}_j - \overline{X})/(s/\sqrt{n_j})$, respectively; $\overline{X}$ and s are the constituent's mean and standard deviation at the well obtained from previous multiple measurements at the well (using at least four sampling periods in a one-year interval). At each sampling period t_j, calculate the cumulative sum S_j:

$$S_j = \text{larger of } \{0, (Z_j - 1) + S_{j-1}\} \quad \text{with } S_0 = 0 \quad\quad (4.2)$$

Plot S_j vs. t_j in the Shewhart-CUSUM (time) control chart. A constituent's concentration is declared "out of control" (that is, exceedingly high) at time period t_j if either $S_j \geq 5$ or $Z_j \geq 4.5$. This would indicate probable well contamination. See Lucas (1982) for further details about the Shewhart-CUSUM control chart.

✓ log trans. first

4.2.2 Trend Detection. Naber et al. (1997) developed a statistical protocol to assess local and global trends in contaminant levels within a groundwater contaminant plume using both parametric and nonparametric methods. In this protocol, temporal trends are examined at three levels: overall temporal trend for the entire site, temporal trends for local regions, and individual temporal trends for single sites. Trends are examined using both parametric and nonparametric methods after transforming the concentration data by taking their logarithms. Seasonal fluctuations that can hinder analysis of long-term trends are also included. The protocol includes the following five steps:

(1)   graphical examination of the groundwater contaminant data to determine the probability
distribution and the presence of seasonal fluctuations

(2)   division of the plume into subregions of interest

(3)   assessment of global trends

(4)   assessment of trends within subregions

(5)   assessment of trends at individual monitoring locations

The graphical methods included in the first step of the protocol include time series plots, which are used to determine whether seasonal fluctuations are present, and probability plots, which are used to determine the probability distribution underlying the groundwater contaminant data. Division of the entire plume into subregions defines areas that are thought to be similar with respect to any criteria. The statistical methods that comprise steps 3 through 5 of the protocol are based on linear regression methods, both parametric (Buscheck and Alcantar, 1995) and nonparametric (Kendall, 1975). For step 5, separate linear regression analyses are performed for each monitoring location.

4.2.3. Geostatistical Methods

4.2.3.1.   Basic concepts. Geostatistics can be defined as a collection of techniques to solve estimation problems that involve geo-referenced spatial variables. Consider a set of groundwater monitoring data, denoted by z_i, in which $i = 1, 2, \ldots, n$, denotes a geo-referenced location. The ith location is defined by a set of coordinates, such as longitude, latitude, and elevation with respect to mean sea level. Given those data, the analyst is often faced with several possible statistical problems: (1) to estimate a variable at a location j (z_j) at which no measurements have taken place, (2) to estimate z integrated over an area or volume of aquifer (say, to determine the mass of contaminant in parts of an aquifer), and (3) to estimate the slope of z at some location in the aquifer. Some of the most interesting problems in groundwater monitoring involve two or more variables whose spatial variation may be interdependent.

It is customary to denote a spatial variable, say, the concentration of a specified contaminant, as the sum of a deterministic mean, m, and a zero-mean random error term (or residual), e. That is, $z_i = m_i + e_i$, in which the index i denotes a location in space identified by a set of coordinates. The spatial random variable z (or "regionalized" variable) may represent the concentration of a monitored contaminant in an aquifer. The mean is, in general, variable in space and represents a trend or the overall level of a regionalized variable. Statistical spatial (and possibly, temporal) dependence is introduced by the residual e. The residual has zero mean and a covariance $C_{ij} = E[e_i \ e_j]$, in which i and j denote two locations in space. It is customary in geostatistics to express the spatial dependence of a regionalized variable in terms of the *variogram* (also called the *semivariogram*) V_{ij}, which is defined as one-half of the variance of the difference of a regionalized variable at locations i and j:

42

$$V_{ij} = \frac{1}{2} \text{Var}[z_i - z_j] \qquad (4.3)$$

The variogram is assumed to be a function of the magnitude of the separation vector **h** between i and j only, and not of its orientation or where i and j might be, in which case the difference $z_i - z_j$ is said to be *intrinsic stationary*. This simplifying assumption can be tested with actual data. If the covariance depends only on the magnitude of the separation vector **h** as well (in which case z is said to possess *second-order stationarity*), the following relationship holds between C_{ij} and V_{ij}: $V_{ij} = \sigma^2 - C_{ij}$, in which σ^2 is the variance of the regionalized z, assumed to be space-independent. A commonly used variogram is the exponential model

$$V_{ij} = V(h) = b(1 - \delta) + A\left(1 - e^{-(h/L)}\right) \qquad (4.4)$$

in which $\delta = 1$ if $h = 0$, else $\delta = 0$, and b is the nugget parameter. The variogram tends to the sill or upper bound $b + A$ when $h \gg L$ and equals 0 when $h = 0$ (see de Marsily, 1986, for more details on the apparent nugget effect); L is called the *integral scale*, a sort of spatial scaling variable.

4.2.3.2. Geostatistical Interpolation. Linear geostatistics estimates z at a specified location $i = 0$ (denoted by z_0) as a weighted sum of the observations z_i, $i = 1$, 2, . . . , n. The estimator z_0^* is then

$$z_0^* = \sum_{i=1}^{n} a_i z_i \qquad (4.5)$$

Assuming that the mean of the regionalized variable z is constant and unknown in the region under consideration, that is, that $E(z) = m$ (where $E(z)$ denotes the expected value of z) and requiring that the estimator z_0^* be unbiased, it follows that the weighting coefficients a_i add up to 1:

$$\sum_{i=1}^{n} a_i = 1 \qquad (4.6)$$

In addition, the error of estimation, $z_0^* - z_0$, is required to have minimum variance, where that variance is expressed by the following equation:

$$\text{Var}[Z_0^* - Z_0] = \text{Var}\left[\sum_{i=1}^{n} a_i Z_i - Z_0\right] = \sum_{i=1}^{n}\sum_{j=1}^{n} a_i a_j C_{ij} - 2\sum_{i=1}^{n} a_i C_{i0} + \sigma^2 \quad (4.8)$$

The minimization of the variance of estimation error subject to constraint (4.6) with respect to the weights a_i leads to the following set of linear equations, from which the weights can be obtained upon solution:

$$\sum_{j=1}^{n} a_j C_{ij} + u = C_{i0} \qquad i = 1, 2, \ldots, n \tag{4.9a}$$

$$\sum_{j=1}^{n} a_j = 1 \tag{4.9b}$$

in which u is a Lagrange multiplier (unknown, must be solved for from [4.9a] and [4.9b]).

The previous equation can be written in terms of the variogram as follows:

$$\sum_{j=1}^{n} a_j V_{ij} + u = V_{i0} \tag{4.10a}$$

$$\sum_{j=1}^{n} a_j = 1 \tag{4.10b}$$

The type of linear geostatistical interpolation just outlined is referred to as *kriging*, after D. G. Krige, a South African mining engineer who developed it. To be more precise, the estimators (4.9) and (4.10) are called *ordinary kriging*. When the mean m is known (in which case it may also be spatially variable also), the systems (4.9a,b) and (4.10a,b) reduce to (4.9a) and (4.10a), respectively, with the Lagrange multiplier u equal to 0, thus giving rise to the *simple kriging estimator*. There are more challenging estimation problems, of course. Those give rise, to cite a few examples, to (1) nonstationary kriging, in which case the regionalized variable may exhibit a nonconstant mean (often required for contaminant data, unless stationary kriging is used in small spatial neighborhoods); (2) co-kriging, when two or more variables are used jointly in geostatistical estimation; and (3) nonlinear geostatistics, which require nonlinear procedures to solve special spatial estimation problems. The reader is referred to the reviews prepared by the Task Committee on Geostatistical Techniques in Geohydrology (ASCE, 1990 a, b) for a thorough exposition of geostatistical techniques. Additional information on geostatistics for hazardous waste applications can be found at http://www.usace.army.mil/inet/usace-docs/eng-tech-ltrs/etl1110-1-175/toc.html. Finally, Reed (2002) compared the performance of six different geostatistical methods for interpolating contaminant concentration data. He found that the performance of different methods was highly dependent on the variability of the concentration data and any preferential sampling of high concentration areas, with a type of kriging called *quantile kriging* showing the least bias from these effects.

4.2.3.3. Variance-based geostatistical methods. A number of monitoring well placement methods rely on geostatistical or probabilistic measures to identify the next location for a new monitoring well or the next sample from an existing well. One of the most common variants of the geostatistical approach in groundwater quality monitoring is the variance-reduction method (Rouhani, 1985), which searches for the number and locations of sampling sites that minimize the variance of estimation error of a contaminant concentration at one or more locations. The search for a set of groundwater sampling locations starts with a number of existing sampling wells to which additional wells from a pool of potential sites are added one at a time. Each additional sampling site is selected on the basis of producing the largest reduction in the variance of estimation error given the existing sampling wells. Sampling sites continue to be added until no further meaningful reductions in the variance of estimation errors can be attained or when marginal gains in statistical accuracy are outweighed by competing criteria, such as budgetary constraints. If there are no originally existing wells, a set of sampling locations must be selected based on hydrogeologic or other appropriate considerations.

Several researchers have relied on minimizing a global function of the estimation error's variance as the basis for selecting monitoring locations (ASCE, 1990a, b). Specifically, they have searched for the best pattern (e.g., square or triangular) and the best density (number of sampling sites per unit area of aquifer) that minimize a global performance criterion, such as the average or maximum variance of estimation of a subsurface contaminant in a region (Olea, 1984). This *global method* is best suited for the selection of the preliminary layout of groundwater monitoring wells, which can be later refined to account for more focused objectives, such as the estimation of contaminant concentrations near water-supply wells or within a high-risk subarea of the aquifer.

A variant of geostatistical search for network design was introduced by Rouhani and Hall (1988). It was used to identify monitoring locations with high uncertainty and high contaminant concentrations. Sampling locations are chosen sequentially in this variant: the next location is chosen to be the area with high uncertainty (as measured by the kriging variance) and high concentration as defined by a threshold constraint. This method could also be used to identify the next sampling location in existing well networks.

4.2.3.4. Concentration-based geostatistical methods.
Variance-based geostatistical methods consider only the spatial covariance structure of the monitoring network but do not consider concentrations of constituents that are being measured. Geostatistical methods can also be used to identify sampling locations that are most critical to achieving good estimates of contaminant concentrations or concentrations of other constituents of interest. Cameron and Hunter (2000) developed several geostatistically based methods to reduce temporal and spatial redundancy. The temporal algorithm consists of constructing a composite temporal variogram, which combines time series data from many wells and uses the variogram to identify appropriate sampling frequencies. . The smallest time interval is identified at which the approximate sill of the variogram is reached, which represents the minimum sampling interval providing essentially uncorrelated temporal data. The sampling frequencies are then adjusted to ensure that the time lags between sampling events do not fall below this minimum interval. The spatial algorithm uses kriging to identify redundant wells that do not need to be sampled. An initial plume map is generated using kriging with all possible well locations. Numerical weights are assigned to each well location to gauge its relative importance to the initial plume map; the weights are

obtained from kriging and are called _global kriging weights_. Subsets of wells with the lowest weights are then removed and the plume map is regenerated. The kriging variance of the new map is calculated and compared with the initial variance to see whether the spatial uncertainty has increased substantially. If not, then additional wells are removed; if so, then the subset of wells are not removed.

4.2.4. Hybrid Hydrogeology–Geostatistics Method

This method is a hybrid that combines geostatistics, geochemistry, hydrogeology, and regulatory knowledge to determine whether or not a well that is part of an existing groundwater monitoring network should continue to be sampled (Tuckfield et al., 2001). The objective of this method is to assess the potential for reducing (1) the number of wells within the selected network, (2) the number of chemical analytes required groundwater per groundwater sample, and (3) the sampling frequency per well within the existing network without compromising the ability to accurately estimate the extent and direction of a plume and the efficacy of the remediation. The Tuckfield et al. (2001) approach is implemented in two phases. The first phase—relevancy, reliability, and regulatory assessment—consists of geochemical and geohydrological evaluations of wells within aquifer zones of interest to ascertain their relevancy, performance, reliability, and regulatory importance. This phase relies on the same principles used in the hydrogeologic approach, reviewed in an earlier section of this chapter, supplemented by regulatory knowledge and policies. The first phase identifies sets of wells that are prospective candidates for deletion from an existing sampling schedule. The effects of fewer data caused by well deletion are evaluated in the second phase—redundancy assessment—using geostatistical analysis. The latter allows the comparison of the predicted spatial concentrations of a target contaminant using all monitoring wells with those obtained by using only a subset of all wells. If the wells in the subset produce a nearly equivalent plume characterization as that forthcoming from all the wells, then the wells not in the subset are classified as redundant and eligible for deletion from the monitoring network. The geostatistical analysis is conducted in a standard fashion, as described in Section 4.2.3.2. In addition to well redundancy, the second phase evaluates the historical trends of contaminant concentrations to determine specific analytes that should continue to be sampled in the wells that remain in the monitoring network.

4.3 Probabilistic Methods

The methods described in the previous sections rely solely on historical data and professional judgment to optimize the monitoring plans. In some cases, it may be desirable to predict the conditions at a site in order to identify the monitoring plans that will be most likely to achieve future objectives. Because of the substantial uncertainty regarding future conditions, these methods typically use probabilistic approaches to identify designs with the greatest predicted reliability. Most of these methods use numerical simulation models to predict future conditions at the site.

4.3.1. *Kalman Filter Methods.* A class of probabilistic approaches uses Kalman filters to identify monitoring well or sampling locations. Herrera et al. (2000) have developed a geostatistical approach that combines a stochastic flow and transport model with a Kalman filter. The flow and transport model is used to compute a space–time contaminant-concentration estimate and its covariance matrix via stochastic simulation. Next, a Kalman filter is used to predict the uncertainty that the concentration estimate would exhibit if concentration data from samples taken from computed locations at different times were used to update the preceding estimate. This is similar to using a space–time kriging method to predict the uncertainty of an estimate, but here, instead of using a space–time variogram obtained from an analysis of concentration data, the variogram is replaced by a space–time covariance matrix, calculated from a transport model via stochastic simulation. A function of the predicted estimate uncertainty is used as a criterion to choose sampling positions and sampling times for the network. This function depends on the objective of the design. For example, it could be the sum of the variance of the concentration estimate at locations close to a drinking-water well during a given year. A sequential procedure is then used to select the space–time sampling point that minimizes the function at each step, and stops when it reaches a predetermined value. For detailed information, see Herrera (1998).

Rizzo et al. (2000, 2001) have developed proprietary software called aLTMOs that uses a Kalman filter to identify both spatial and temporal sampling plans. It uses a method to overcome difficulties that can arise with Kalman filters in estimating and propagating the space–time covariance matrix. The method can be used to correct predictions from analytical or numerical transport models when new data are obtained and to identify sampling locations that are most important for ensuring accurate future results. New monitoring well locations or sampling locations in existing networks are selected to increase the information value of the monitoring network and to reduce estimates of uncertainty for the purposes of satisfying one or more monitoring objectives. Data from the existing monitoring network are used to estimate concentrations at various locations, as well as the estimation uncertainties at those points. If, for example, the goal in installing a new monitoring well is to increase the confidence in the location of a specified contour level, then the location of the contour level as well as a confidence band in its location can be plotted. Because it is often important to reduce the risk of false-negatives, values of variance within the confidence band are examined to identify where variance is greatest (i.e., the location where an additional monitoring well would cause the greatest reduction in risk that the contour is incorrectly located). Figure 5.5, in the next chapter, shows an example visualizing the confidence band around an action-level contour, as well as estimates of variance within the band. The best location for a new monitoring well, given that the purpose of the well is to best locate the action-level contour, is where the variance within the band is highest.

Kalman filter methods can also be used to identify sampling locations and frequencies in existing monitoring networks. Rizzo et al. (2000) have used their aLTMOs system to identify both spatial and temporal sampling plans in existing networks. The method can be used to correct predictions from analytical or numerical transport models when new data are obtained and to identify sampling locations that are most important for ensuring accurate future predictions. These locations can then be retained in the sampling plan.

4.3.2. Probabilistic Simulation. Other probabilistic approaches use numerical simulation to identify the monitoring well locations that are most likely to detect contaminant plumes as they move down-gradient over time. Massmann and Freeze (1987) presented a guide for landfill monitoring design that uses conditional simulation to calculate the probability of contaminant detection. In conditional simulation, uncertain aquifer properties, such as hydraulic conductivity are generated synthetically. Values for a spatially distributed aquifer property are conditioned on actual field measurements (Yeh, 1992). For each generated field (also called a realization), a numerical model is implemented to describe the transport of contaminants in the aquifer. Thus, many realizations (typically thousands) and the associated numerical simulations yield a sample of spatial and temporal contaminant concentration values in the aquifer. This is a Monte Carlo simulation approach for the development of concentration probability distributions based on synthetic realizations of aquifer properties. For any given arrangement of monitoring wells and sampling frequency, the simulation approach yields important information, such as the probability that a contaminant plume might go undetected (i.e., the probability of a false- negative). Ultimately, probabilistic simulations identify efficient sets of plausible groundwater quality sampling sites (i.e., ones that ensure plume detection at critical points in a timely fashion without requiring an excessive number of sampling points).

Another example of a probabilistic method is presented by Ely et al. (2000) and Hill et al. (2000). The method determines the decrease or increase in prediction uncertainty produced by omitting or adding observations, using linear, second-moment analysis. The statistic developed for the analysis is called the *observation-predictions statistic* (OSP), and is most accurately thought of as a measure of leverage instead of influence; that is, the OSP statistic is calculated by omitting or adding observations to the first-order, second-moment equations and is not a method, such as jack-knife methods, in which observations are removed or added one at a time and the regression is repeated. As such, when evaluating potential new observations, the OSP method does not depend on the observed value, but only on the type and location of the potential measurement, and the likely accuracy of the measurement. In the cited references the OSP method is used to evaluate the importance of sets of one or more existing observation locations to predictions of interest. As noted, it can equally well be used to evaluate the importance to predictions of possible new observations. The authors describe a steady-state model, but the method is directly applicable to transient models in which system conditions for data collection differ from those that are likely to occur during the time frame of the prediction of interest.

4.3.3. Hierarchical Method. Hierarchical monitoring provides yet another option among the probabilistic methods for long-term groundwater quality monitoring. The method, by its sequential structure, is a mixture of probabilistic analysis and optimization. For example, Scheibe and Lettenmaier (1989) considered three levels in the monitoring decision problem: (1) contaminant, geographic, and hydrogeologic reconnaissance; (2) estimation of the probability of water-supply well contamination; and (3) selection of sampling sites. Their objective function was to minimize the aggregate exposure risk of population centers served by groundwater. The aggregate exposure risk was defined as the sum of the product of well-contamination probability times the population served by the well, where the sum is carried out over all water-supply wells in the study area. Scheibe and Lettenmaier (1989) illustrated the hierarchical method with regional monitoring of pesticides in groundwater.

4.4 Mathematical Optimization Methods

Mathematical optimization methods are numerical search algorithms that can be used to automatically search for optimal monitoring designs that satisfy user-specified monitoring objectives and constraints. These methods are most appropriate for use when the number of possible designs makes manual search cumbersome, such as at sites with any of the following traits: (1) numerous existing or potential monitoring wells, (2) numerous constituents that need to be considered in sampling design, or (3) different types of samples with varying levels of accuracy and cost (such as indicator samples or sensor data). Under these conditions, the identification of optimal sampling plans with heuristic or enumerative schemes may become prohibitively time consuming. Consider, for example, a situation in which there are n wells and at each well there are m constituents that could be measured in any monitoring period. The number of possible sampling plans in this instance is $2^{n \times m}$. If, say, $n = 10$ wells and $m = 3$ constituents, the number of possible sampling plans would be 2^{30}, or slightly over 1 billion plans. Even if 99% of these plans could be eliminated using professional judgment and each plan could be evaluated in 1 s, it would still take 120 days to evaluate all of the 100 million remaining plans. Mathematical optimization methods offer an automated approach for identifying optimal sampling plans in a cost-effective and expeditious manner.

To use optimization methods, the stakeholders involved in the site identify the desired monitoring objectives. The objectives are translated into an objective function that quantifies the worth (or cost) of various candidate sampling plans. Many types of objective functions have been used in past work related to groundwater quality monitoring, including the minimization of monitoring cost (Reed et al., 2000, 2001a, 2001b), maximization of the accuracy of concentration predictions (Loáiciga, 1989), maximization of plume-detection probability (Meyer and Brill, 1988; Meyer et al., 1994; Storck et al., 1997), maximization of monitoring coverage (Hudak et al., 1995), minimization of the error in interpolated concentrations (Reed et al., 2001b), and many others. Any objectives can be included as long as there is some way of quantifying performance of candidate monitoring plans using that objective (such as a cost function that describes sampling costs or a plume interpolation method or simulation model that estimates contaminant concentrations).

A few studies (Meyer et al., 1994; Cieniawski et al., 1995; Reed et al., 2001a, 2001b; Reed, 2002) have identified optimal sampling plans for multiple objectives, where each plan represents the best tradeoff between the objectives. Figure 4.3 shows an example of an optimal tradeoff curve between the conflicting objectives of

49

minimizing relative cost and minimizing relative error in interpolated concentrations. Each point on the curve represents one possible sampling design that is optimal for a given level of sampling cost. Any level of monitoring cost is associated with a set of optimal monitoring locations that is defined by a number of wells and their locations. For example, one point may be associated with 10 sampling locations, whereas another point with higher cost but improved accuracy would involve a much larger number of sampling locations.

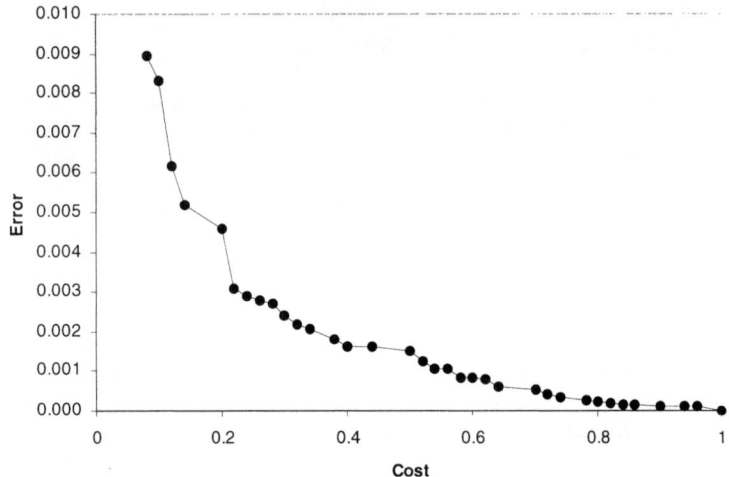

Figure 4.3. Tradeoffs between sampling cost and error for various sampling designs. Each point represents one possible sampling design that has been optimized for a specific tradeoff.

Many objectives, such as maximizing accuracy of concentration predictions or maximizing plume-detection probability, require that the effectiveness of monitoring plans at meeting these objectives be evaluated. To perform this evaluation, optimization models often use statistical techniques for spatial and/or temporal interpolation or simulation of probable plume locations to ensure that contaminants will be detected. Meyer and Brill (1988) combined the plume conditional simulation approach of Delhomme (1979) with optimization to identify well locations that maximize the probability of plume detection while minimizing the sampling cost. Meyer et al. (1994) expanded the approach to consider the area of an aquifer that was contaminated before initial detection. Hudak and Loaiciga (1993) used a mass transport model to define probable impact zones down-gradient of a landfill. In turn, the impact zones were used to establish weights for candidate monitoring sites. Reed et al. (2000) used ordinary kriging, inverse-distance weighting, and a hybrid of the two approaches to evaluate how accurately the total amount of contaminant mass in the subsurface could be interpolated using candidate sampling plans. Reed et al. (2001b) used a variant of a nonlinear least squares interpolation method developed by Barry and Sposito (1990). The focus of both studies (Reed et al. 2000, 2001b) is on well redundancy, where the accuracy of candidate sampling plans was evaluated relative to the accuracy of interpolation using all available wells.

Once appropriate monitoring objectives have been identified, any relevant constraints must be included. If only one objective is being included in the analysis (e.g., minimizing cost), then a second objective, such as minimizing error, can be included as a fixed constraint specifying that the error cannot exceed a maximum acceptable value. Other site-specific constraints can also be included in the optimization model. Examples are down-gradient sentinel wells that must be sampled or regions of the site where monitoring wells cannot be located, budgetary constraints, and minimum number of monitoring wells. In some instances it is computationally advantageous to transform constrained problems into equivalent unconstrained problems, which are often easier to solve. Penalty functions and Lagrange multipliers are two techniques that are used for that purpose (e.g., see Hillier and Lieberman, 1995).

Once an appropriate optimization model (which consists of objectives and constraints) is formulated, many optimization methods exist that can search for optimal monitoring plans (sampling locations, frequencies, types of samples, etc.) that achieve the monitoring objectives and satisfy constraints. These methods would be linear programming, nonlinear programming, integer programming, mixed integer programming, simulated annealing, and genetic algorithms, among others. The most appropriate method for a given application depends on the form of the objectives and constraints. Given the discrete form of most monitoring optimization problems, linear and nonlinear programming are not usually applicable (see references such as Revelle et al., 1997, for details on these methods). A brief overview of the most widely used network-design optimization methods follows, along with references for further information.

4.4.1. Integer Programming.
Integer programming has arguably been the most widely used approach to solving monitoring design problems with mathematical optimization (see Meyer and Brill, 1988; Loaiciga, 1989; Hudak and Loaiciga, 1993; Wagner, 1995; Storck et al., 1997; Mahar and Datta, 1997). Integer programming can be applied to optimization problems where the decision variables (such as whether or not to sample a particular location) take on only integer values. For example, the decision of whether to sample a particular location is usually represented with a variable that has the value 1 if the location is sampled and 0 if not; this special case is sometimes called a binary problem. Integer programming problems are usually solved using branch-and-bound methods, in which a tree of possible solution values is set up and each branch is examined and pruned until the optimal solution is found. (See Revelle et al., 1997, for an introduction to the method.) For such cases, practical computing limitations may restrict integer programming to problems having a relatively small number of decision variables. Appendix B contains a formulation example of an integer programming problem for network design.

51

4.4.2. Simulated Annealing. Several heuristic, probabilistic, mathematical optimization algorithms can be used to search for optimal monitoring designs, of which simulated annealing and genetic algorithms are prime examples (genetic algorithms are described in the next subsection). The main advantages of these algorithms are that (1) they can be applied to any type of optimization problem; (2) they can easily be combined with existing simulation or plume interpolation routines without major recoding; (3) they are particularly effective for rapidly finding good, though not necessarily optimal, solutions to very large optimization problems, substantially larger than integer programming can handle; and (4) they can be applied by persons other than experts, although experienced users can obtain significant improvements in performance and results.

Dougherty and Marryott (1991), Marryott et al. (1993), and Rizzo and Dougherty (1996) applied simulated annealing methods to groundwater remediation design, but their approach can easily be adapted to monitoring design. The method of simulated annealing originates from the analogy between the optimization of a function and the behavior of thermodynamic systems. Metropolis et al. (1953) introduced an algorithm that incorporates the idea of choosing the most probable system behavior in numerical calculations. The key to simulated annealing is to slowly lower the "temperature" through incremental steps such that the system will "anneal" from a state of random order to a state of high order, which results in a (near) global minimum objective. This annealing procedure is implemented by optimizing (e.g., minimizing a cost function) over a series of iterations, or "temperature" increments, using what has become known as the *Metropolis algorithm*. For each temperature (T) increment, a reference configuration (set of sampling wells) and a trial configuration are selected. A flow and transport simulation is carried out, and the resulting objective or cost function, C_{ost}, is computed. If the difference between the current cost function and the previously accepted reference cost function, ΔC_{ost}, is less than or equal to 0 (i.e., the new sampling network configuration reduces cost), the new configuration is accepted. If, however, $\Delta C_{ost} > 0$, the probability that the new configuration will be accepted is

$$P(\Delta C_{ost}) = \exp\left(\frac{-\Delta C_{ost}}{T}\right) \qquad (4.11)$$

This probability can then be compared with a pseudorandom number between 0 and 1. If the random number is less than $P(\Delta C_{ost})$, the new configuration is accepted; if not, the original configuration is used to start the next iteration at the same temperature step. Periodically, the temperature is reduced to restrict the tendency to accept higher-cost candidates. Thus, the Metropolis algorithm always accepts improved sampling network configurations, and probabilistically rejects deteriorated configurations. Accepting unimproved configurations enables the system to escape local minima. In theory, simulated annealing will find the global minimum of the objective function, although in practice the procedure is terminated at or near global minimum.

The basic simulated annealing algorithm requires that five components be specified (Rizzo and Dougherty, 1996). Three of these are often application-independent: (1) the candidate-generation method (which generates new candidate solutions that are examined for optimality), (2) the annealing schedule (which determines which candidate solutions are acceptable and which are not), and (3) the stopping criterion. In some cases, however, significant performance enhancements can be obtained by tuning these components. The remaining two components, the cost function and data representation for the system being optimized, are problem-dependent. In the hands of experienced users, significant improvements in computational performance can be obtained by manipulating the probability distributions that are used within the candidate-generation method.

4.4.3. Genetic Algorithms. Another optimization approach that has been implemented for monitoring design is genetic algorithms (Wagner, 1995; Reed et al., 2000, 2001a, b). Genetic algorithms, which were developed by Holland (1975), simulate the mechanisms of natural selection in searching for optimal solutions. In using genetic algorithms, "strings" are formed that are digital representations (binary or decimal) of different decisions. For monitoring design, each string (called a *chromosome*) would represent one sampling plan. In a binary representation, a string could be 01110, which could represent not sampling from the first well, sampling from the second through fourth wells, and not sampling from the fifth well.

The genetic algorithm starts with a random initial "population" of strings and evolves them using three basic operations: (1) selection, (2) crossover (mating), and (3) mutation (see Goldberg, 1989, for reference). Each string is then evaluated based on its performance (fitness) with respect to the objective function and constraints. Using this fitness value, strings are selected to enter the mating population using one of a number of selection approaches. The crossover or mating operation involves exchanging genetic information between pairs in the mating population at randomly selected locations on their chromosomes (strings). Finally, mutation occurs when the bits of a string, called *genes*, are randomly changed with a specified probability. By repeating these three basic operations for a number of generations (iterations of the model), the performance of the population continues to improve.

The general theory behind this process is that strings with high fitness values contain information chunks ("building blocks") that are important to optimizing the objective function. By exchanging important building blocks between two strings that perform well, the genetic algorithm attempts to produce children strings that contain the important building blocks from both parents and, therefore, perform even better than the parent strings. In this way, genetic algorithms use Darwin's survival of the fittest theory to search for the best solutions. It is through this process of assembling strings with important building blocks that an optimal solution is found. The algorithm is halted when the best string in the population takes over, usually by representing more than a specified percentage (e.g., 90%) of the strings in the population.

The major advantage of using genetic algorithms, beyond those discussed in the previous section, is that they search from an entire set of possible solutions in the population. This feature makes them particularly well suited for implementation on parallel computers. Moreover, it allows genetic algorithms to find an entire range of solutions that satisfy multiple objectives within a single run (as shown in Figure 4.3). In principle, tradeoff curves such as those shown in Figure 4.3 can usually be obtained using any other single-objective optimization approach by, for example, minimizing cost as the objective and setting a maximum allowed error as a constraint. The tradeoff curve could then be generated by varying the allowed error. Although such an approach can be done, it is often far more difficult and time-consuming than using a multiobjective genetic algorithm, particularly for complex problems such as monitoring design.

Appendix 4.A

Method for Removing Monthly Fluctuations from Time Series Data. There are several techniques for removing seasonal components in time series data, either time- or frequency-domain based. One widely used to remove known monthly fluctuations is as follows. Suppose that there are $12 \times N$ months of concentration data collected during N years. The mean concentration of each of the 12 months is calculated from the N years of data, and denoted by $\overline{X}_i$, i = 1, 2, . . . , 12. The grand mean of all the concentration data ($12 \times N$ observations) is also calculated; let it be denoted by $\overline{X}$. The adjusted concentration data Z_{ij}, for the ith month, $i = 1, 2, . . . , 12$, and j^{th} year, $j = 1, 2, . . . , N$, is calculated with the following equation:

$$Z_{ij} = X_{ij} - \overline{X}_i + \overline{X}$$

The variable Z_{ij} represents monthly concentration with the seasonal component removed.

Appendix 4.B

An Integer Programming Example of Monitoring Network Design. In the interest of specificity, here is an example of a mathematical optimization model for network design. A simplified version of the mixed-integer programming formulation advanced by Hudak et al. (1995) is used. Consider Figure 4.1, in which there are three layers (upper and lower sand layers and a glacial till). The overall goal is to find the optimal groundwater sampling well locations in the threatened aquifer system posed. Specifically, this is translated to an objective function (Z) that seeks to maximize the information content acquired from groundwater samples. Let us introduce the following notation:

$j = 1, 2, . . . , J_k$ areal (map-view) index of potential well site in layer k;
$k, m = 1, 2, . . .$, K layer index (the index m is used for vertical interaction constraints, see below);
J_k = set of potential well sites in layer k;
K = number of layers ($K = 3$ in this example);

w_{jk} = weight for node j, layer k (these weights must be specified by the analyst based on a variety of risk-assessment considerations, see Hudak et al., 1995);

$x_{jk} = 1$ if a well is installed at site j, layer k; else $x_{jk} = 0$ (x_{jk} is a decision variable; the notation x_{jm} is used for vertical interaction constraints, see below);

$v_{jk} = 1$ if a well at site j and layer k is part of a well cluster (i.e., wells located side-by-side but screened within different aquifer layers); else $v_{jk} = 0$ (v_{jk} is a decision variable);

$P_k(\text{min})$ = minimum number of wells to be located in layer k;

$P_k(\text{max})$ = maximum number of wells to be located in layer k;

P = total number of wells to be located;

ω = weight varied in the interval [0, 1].

The objective function of this well-location problem is

$$\text{Max } Z = \omega \sum_{k=1}^{K} \sum_{j=1}^{J_k} w_{jk} x_{jk} + (1 - \omega) \sum_{k=1}^{K} \sum_{j=1}^{J_k} w_{jk} v_{jk} \tag{B.1}$$

The first term in the right-hand side of equation (B.1) is the sum of weights for all potential well sites placed in the layered aquifer. If a well is located at a site, say (j, k), it contributes the corresponding weight value (w_{jk}) to the objective function. The second term in the equation (B.1), referred to as the *vertical interaction* term, is the sum of the weights for nodes that are part of well clusters. The relative values for the two terms in the objective function (B.1) depend on the partitioning in emphasis established by the weight ω. Equal emphasis is assigned by setting $\omega = 1/2$.

The nodal weights w_{jk} account for (1) hydrogeologic, (2) human exposure, and (3) covering factors. Hydrogeologic factors are proportional to the seepage velocity of ground water and inversely proportional to the distance of a monitoring well to a contaminant plume. A plume's geometry is estimated by an advective envelope that ensues from a pollution source. Exposure factors are proportional to the human population served by a water supply well (which is also used as a monitoring location) and inversely proportional to the distance of the well to a contaminant plume. Covering factors measure the marginal information gain obtained from placing a well at a particular location relative to other possible locations. See Hudak et al. (1995) for details on the procedure for calculating nodal weights.

The following constraints complete the formulation of the network design problem:

$$\sum_{j=1}^{J_k} x_{jk} \geq P_k(\text{min}) \qquad k = 1, 2, \dots, K \tag{B.2}$$

Constraint (B.2) sets a lower bound to the number of wells in each layer.

$$\sum_{j=1}^{J_k} x_{jk} \leq P_k(\text{max}) \qquad k = 1, 2, \dots, K \tag{B.3}$$

Equation (B.3) limits the maximum number of wells in the kth layer.

$$\sum_{k=1}^{K} \sum_{j=1}^{J_k} x_{jk} = P \tag{B.4}$$

Equation (B.4) enforces the required total number of sampling wells in the aquifer.

$$v_{jk} \leq x_{jk} \quad j = 1, 2, \ldots, J_k; \ k = 1, 2, \ldots, K \tag{B.5}$$

Constraint (B.5) is a vertical interaction constraint, which defines conditions necessary for well clustering: it specifies that a potential well location cannot be part of a well cluster unless a well is sited at that location.

$$v_{jk} \leq \left(\sum_{m=1}^{K} x_{jm} \right) - 1 \quad j = 1, 2, \ldots, J_k; \quad k = 1, 2, \ldots, K \tag{B.6}$$

Constraint (B.6) defines a well cluster as two or more wells at a single areal location j. Both x_{jk} and v_{jk} are binary integers. However, for a given j and k, v_{jk} can only equal 1 if a well is screened at site (j, k) (i.e., $x_{jk} = 1$) and if one or more wells are screened above or below layer k at the areal location j. Lastly, binary integers conditions are imposed on model decision variables:

$$x_{jk} = (0, 1) \quad j = 1, 2, \ldots, J_k; \quad k = 1, 2, \ldots, K \tag{B.7}$$

$$v_{jk} = (0, 1) \quad j = 1, 2, \ldots, J_k; \quad k = 1, 2, \ldots, K \tag{B.8}$$

Solution of the mathematical optimization problem embodied in equations (B.1) through (B.8) yields the optimal well sampling locations and well clusters in the threatened aquifer. The optimization problem so defined can be solved with commercial solvers (e.g., Excel Solver, LINGO, CPLEX).

Chapter 5: Field Studies of Long-Term Monitoring Design

The main body of this chapter provides summaries of field studies in which long-term monitoring (LTM) design methods have been applied at actual sites. Table 5.1 summarizes these field studies, classifying them by the methods used for optimization and by the monitoring objectives. Note that the methods listed in Table 5.1 are not the only methods that can be used for LTM design; they are simply the methods that have been used in these cases. Only case studies in which the methods have been implemented at actual sites are included in this chapter, whereas Chapter 4 described a variety of other methods that may not yet have been applied at field sites. This chapter includes only field studies that have not been published previously. Table 5.2 summarizes a number of other LTM field studies that have been published previously.

The field studies compiled in this chapter illustrate a number of important findings for LTM design. These findings are summarized in the following paragraphs.

5.1. Relationship Between Site Remedies and Performance Monitoring

The long-term monitoring of groundwater systems is conducted to provide the data needed to ensure the risks to human health and the environment are being properly managed in accordance with site restoration plans. The restoration plans are site specific and are a function of the amount of data available. Although the regulatory and permitting processes usually require that the remedy first be approved in a record of decision prior to the initiation of monitoring, the remedy and the monitoring *are inextricably related*. The performance of the remedy can be evaluated only by monitoring for parameters that indicate whether that particular type of performance is being met. As data continue to be collected and site knowledge increases, the selected site remedy may change, as should the corresponding performance-monitoring scheme. A common example of this dynamic is the discovery that remedy/monitoring scenarios selected to verify plume stability (e.g., containment) lead to nearly perpetual operation and prohibitive life cycle costs. This results in a change in the remediation objective to something other than containment (i.e., mass removal), so that the time horizon for cleanup is shortened, which in turn necessitates a new performance-monitoring plan. The importance of designing monitoring plans that are compatible with the site remedies cannot be overemphasized.

Optimization of the performance-monitoring plans can be accomplished using a variety of approaches. The selection of an appropriate method involves numerous criteria, the most important of which are the amount and type of available data, and the site-specific long-term performance objectives (see Table 4.2). Typical questions that monitoring networks should address are as follows: (1) where is the plume? (2) what are the peak concentrations and the distribution of mass within the plume? (3) what do we estimate the peak concentrations will be at selected points or regions of interest? (4) what is the direction of flow? (5) is the plume stable, shrinking, or expanding? Each estimate has an associated uncertainty, and the uncertainties in the error of estimation can be used as decision criteria.

Several approaches have been developed for identifying suitable monitoring plans. As mentioned in the previous chapter, most of the early work in monitoring design focused on methods for siting new monitoring wells for the purpose of site and plume characterization at landfills or hazardous waste sites (monitoring objectives 1 and 2 of Table 4.2). Much of the more recent LTM optimization work, presented in the following case studies, has been developed to identify sampling plans that minimize the spatial and temporal redundancies in existing monitoring networks.

5.2. Plume Shrinkage vs. Stability

Seven of the eight case studies presented in Table 5.1 are associated with monitoring objective 4 (verify plume shrinkage) and objective 5 (verify plume stability). In selecting a monitoring design methodology, the distinction between these two objectives is important because objective 4 implies that the plume is nonstationary (i.e., a plume that is moving and has a time-varying distribution of contaminants within the plume) and objective 5 implies that the plume is relatively stable and can be assumed to be stationary in time.

When the rate of plume change is infinitesimally small, steady-state analysis methods can be brought to bear. In these circumstances, diagnostic methods of data analysis mentioned in Chapter 4 (Sections 4.1.2 and 4.2.3) can be used, and complex analyses of time-varying conditions are not needed. In this scenario, groundwater monitoring is similar to that of soil contamination monitoring, and numerous methods are applicable based on either interpolation theory (e.g., polynomials and Delaunay triangulation) and statistical methods (e.g., trend analysis and linear regression) to verify that there is no temporal trend or spatial geostatistical methods, such as kriging. All of these approaches can be expressed as taking samples to minimize the error (or the integral of error) over the region of interest. By contrast, when the plume is changing rapidly, the dynamics of the plume become important and a physically based probabilistic approach (as described in Section 4.3) is better than a purely spatial approach. In either case, the amount of data influences which analysis methods are *appropriately* used, not simply which ones *can* be used.

5.2.1 Plume Stability. If the plume and/or the source region has a stabilized distribution of contaminants that are decaying, the main performance-monitoring problems are (1) ensuring the distribution is stable, (2) estimating the distribution of contaminants within the source zone, and (3) estimating the net decay rate for mass in the plume. Note that monitored natural attenuation (MNA) is often instituted in the context of a "stabilized plume," in which a balance exists among advection, dispersion, and reaction such that the shape and distribution of contaminants are not changing in time, while the magnitude of the contamination is decreasing.

5.2.2. Plume Shrinkage. Plume monitoring for *cleanup* (monitoring objective 4 of Table 4.2) requires that the amount of mass in the subsurface be adequately characterized, i.e., to ensure that peak concentrations or concentrations encountered at points of compliance decrease or reach specified values. Because the plume is actively manipulated, either hydraulically or chemically, the remedy creates transient behavior of the contaminant distribution within the plume. Thus, the optimal monitoring plan needed to estimate the concentration (or total mass) over the region of interest must be planned with this time dependency in mind. Often the active cleanup period will be followed by MNA. When MNA is applied to a nonstationary plume, a spatially fixed monitoring network is generally not optimal over the span of a monitoring operation and will need to be shifted downstream in response to plume movement and redistribution of contaminants. This is analogous to "plume chasing" in the discovery and site characterization phase of a project. Interestingly, the MNA problem is much more challenging than this plume discovery and characterization phase, because MNA is usually monitored using a number of analytes and some of these will be in stabilized distributions whereas others used for performance monitoring are simultaneously in nonstationary plumes (e.g., dechlorination reduces solvent concentrations but acts as an ongoing source for a mobile chloride plume).

Table 5.1. Summary of the case studies of Chapter 5 based on methods and monitoring objectives

Authors	Location and Optimization Method	Monitoring Objectives	Results and Savings
Hudak	Tarrant County, Texas Mathematical optimization methods (4.4)	Release detection	Used modified version of the Location Set Covering Model, solved by LINDO to identify 1) deficiencies in existing network design and 2) the minimum number of borehole locations needed to detect contaminant plumes beneath a landfill.
Tomasko	Superfund site 4.1.1 – Hydrogeologic method 4.1.2 – Cost-effective sampling	Verify plume stability or shrinkage	Used DQOs and a key sampling parameter scoring system to provide a quantitative estimate of the utility of sampling locations for post-remediation monitoring. Reduced the initial monitoring plan consisting of 80 wells and 15 springs to 6 wells and 7 springs.
Tuckfield, Shine, Hiergesell, Denham, Beardsley & Reboul	DOE Savannah River Site 4.2.4 Hybrid hydrogeology-geostatistics method	Verify plume stability or shrinkage	Identified redundancy in existing monitoring well networks and recommended elimination of analytes from the sampling schedule.
Aziz, Newell, Rifai, Ling & Gonzales	McClellan AFB 4.1.2 – Monitoring and Remediation Optimization System (MAROS)	Verify plume stability or shrinkage	Used MAROS (parametric and nonparameteric trend analysis; Delauney triangulation) for redundancy analysis and sampling frequency optimization to identify redundancy in existing monitoring well network, improved future monitoring plan sampling frequency.
Lillys, Sullivan & Morgan	NAS Fort Worth, JRB, Fort Worth, TX 4.2.2 – Trend detection 4.2.3 – Geostatistical method	Verify plume stability or shrinkage	Used geostatistical methods for spatial and temporal redundancy to reduce the existing network design by 60% (requires 72 sampling locations as compared to 193 in 1999)
Cameron & Hunter	Massachusetts Military Reserve (MMR) 4.2.2 – Trend detection 4.2.3.4 – Concentration-based geostatistical method	Verify plume stability or shrinkage	Identified 20% of monitoring wells as spatially redundant and relaxed sampling frequencies from quarterly to annually and in some cases to once every 5 years.
Rizzo & Dougherty	Army SSCOM, Natick MA 4.2.2 – Trend detection 4.2.3 – Geostatistical method 4.3.1 – Kalman Filter method	Verify plume stability or shrinkage; verify or correct site models	Used Bayesian filter and site flow and transport model to verify/correct site model and perform uncertainty forecasts. Also used geostatistical methods for spatial and temporal redundancy analysis to monitor containment along a property boundary with same amount of uncertainty and indicate where an additional monitoring well can reduce uncertainty and maximize the value of data.
Herrera, Guarnaccia & Pinder	Superfund site 4.3.1 – Kalman filter method	Verify plume stability or shrinkage	Used transport model and Kalman filter for spatial and temporal redundancy analysis to check containment on part of the property boundary with same amount of uncertainty, the initial monitoring plan consisting of 322 samples from 62 wells was reduced to a plan consisting of 40 samples from 23 wells.

Table 5.2. Summary of previously published case studies

Authors	Location and Optimization Method	Results and Savings
Klawitter and Barry (1999)	Naval Air Station, Brunswick, ME. Mostly solvents and pump-and-treat with UV/oxidation. Long-term monitoring.	Intially 36 MWs sampled quarterly at $550K/yr. Geostatistics and DQO led to 22 MWs at reduced frequency, data gap MWs, and savings of about $300K/yr.
Cronce, Gass, and Trentacoste (1999)	80-acre Naval Industrial Reserve Ordnance Plant, Fridley, MN. TCE and 6-well pump-and-treat with air stripping. Long-term monitoring.	Cost of assessment $40K. Savings of about 20% using SmartSite approach.
Rob Greenwald (1999)	Applied hydraulic optimization techniques at two DOD sites and one industrial facility.	Two of the three sites suggested savings of millions of dollars (net present value) over 20 years, relative to existing designs.
Hansen (1999)	Badger Army Ammunition Plant. Long-term monitoring.	Initially 220 wells. Reduced to 170 wells, reduced number of analytes, reduced frequency. Savings of $400K/yr from $1,400K/yr in LTM. Methods involve data visualization and "common sense procedures."
Nabor et al. (1999)	Formerly used defense site. BTEX and MTBE, with air sparging and natural attenuation.	Initially 54 MWs. Reduced to 49 MWs. 10% savings in sampling and lab costs using "redundancy analysis."
Hassig et al. (1999)	Pantex (DOE) Site, extensive list of contaminants. Using DQO process to develop basis for decisions.	Initially 70 MWs on quarterly or semi-annual schedule. Claim reduction of $400K/yr using unspecified "decision logic."
Singh and Harvey (1999)	McClellan AFB, soil vapor extraction.	Save $30K/yr in analytical costs using decision tree method.
Hunter (1999) Cameron (1999)	Wurtsmith AFB. TCE and *trans*-1,2-DCE with pump-and-treat and air stripping.	Reduced frequency 75% using time series analysis and LLNL decision tree method.
Tuckfield and Ridley (1999)	LLNL.	$200K/yr (25 to 40%) savings in monitoring. Cost of analysis $85K using decision tree ("cost effective sampling") method.
Johnson, Ridley, Tuckfield, and Anderson (1996)	Savannah River Site (DOE), Area F. Nine contaminants.	Initially 89 MWs on quarterly schedule. Reduced to 35MWs quarterly, 51 MWs semiannually, 3 MWs annually, for estimated $177K/yr lab savings.
Ridley and Johnson [1996]	LLNL Main Site and Site 300. TCE, with pump-and-treat and others.	Initially (1992 data) for Main Site/Site 300: 212/297 qrtly MWs, 77/0 semiannu. MWs, and 7/26 annu. MWs. Decision tree ("cost effective sampling") led to 81/180 qrtly, 65/117 semiannu., and 150/134 annu. MWs. Savings of $230K/yr Main Site and $160K/yr Site 300.

5.3. Case Studies

Monitoring Groundwater Beneath A Solid Waste Landfill in Tarrant County, Texas

Paul F. Hudak[1], University of North Texas

A configuration of nested monitoring wells was designed for a 32-ha solid waste landfill in Tarrant County, Texas (Figure 5.1). Positioned above a sand aquifer and between two ephemeral streams, the landfill is a potential source of groundwater contamination (Hudak, 1998). A 0.5 m to 1.0 m clay layer lines the bottom of the landfill. The unconfined aquifer, averaging 9 m in saturated thickness, overlies shale bedrock. An existing network of seven down gradient monitoring wells is screened from 6 m to 8 m across the water table (Figure 5.1).

MODFLOW (McDonald and Harbaugh, 1988) and MT3D (Zheng, 1990) simulated a subset of potential contaminant plumes emerging near the down gradient boundary of the landfill. Ten potential monitoring loci, each perpendicular to (non uniform) groundwater flow, were established in a buffer zone down gradient of the landfill. Five potential monitoring loci were located in each of two model layers. Although the models simulated only one geologic unit, multiple layers were used to permit nested monitoring.

For each monitoring locus, the maximum width (W) attained by the narrowest simulated plume was compared with the total length (L) of the locus. Monitoring sites were identified on the locus with the highest W/L ratio.

A modified version of the Location Set Covering Model (Toregas et al., 1971), solved by LINDO (Schrage, 1991), identified the locations of the minimum number of boreholes needed to detect all of the simulated contaminant plumes. The resulting network consisted of 10 boreholes with vertically nested intakes (two intakes in each borehole). By comparison, the existing network detected only 44.9% of the simulated contaminant plumes.

The existing network (Figure 5.1) illustrates three fairly common deficiencies: (1) an insufficient number of monitoring wells, (2) sampling locations that are too close to the landfill, and (3) erratic spacing between wells. Moreover, existing detection networks often lack nesting, which could enable detecting contaminant plumes at different depth intervals, documenting the vertical distribution of contamination, and tracking the progress of corrective action measures.

A remedy was proposed to address the deficiencies outlined above. Four well nests were added between existing wells on the landfill's southern downgradient boundary. The well nests were located approximately 50 m downgradient from the landfill. No augmentation was necessary along the eastern downgradient boundary. Overall, results of this study illustrate a practical need for structured approaches to designing detection networks in aquifers beneath landfills.

---

[1] Dr. Paul F. Hudak, Department of Geography and Environmental Science Program, University of North Texas, P.O. Box 305279, Denton, TX 76203-5279, USA, e-mail: hudak@unt.edu.

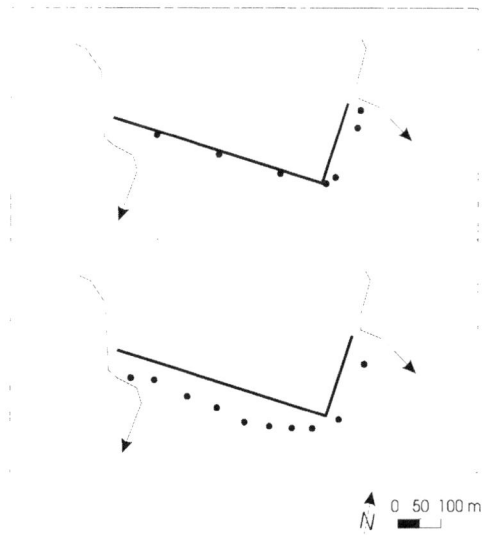

Figure 5.1. Plan view of landfill's down gradient boundary (bold), and monitoring wells (dots) in (top) existing network and (bottom) derived network. Lines with arrows depict ephemeral streams.

Optimizing a Groundwater Monitoring System for Post-Remediation Sampling

David Tomasko[2] and Gustavious P.Williams, Argonne National Laboratory

As an integral part of the Remedial Investigation/Feasibility Study (RI/FS) process, groundwater and springs are sampled as part of site characterization. Following completion of the RI/FS process, post-remediation monitoring is frequently desired to evaluate the effectiveness of the remediation activities. By implementing a seven-step data quality objective (DQO) process, constraints on the monitoring system can be developed to ensure the adequacy of the monitoring system and sampling protocol. Although the same wells and springs used in site characterization can be used for post-remediation monitoring, the system may not be optimal, as required under the DQO process. That is, wells and springs may be providing redundant information, or the information being provided may not be useful for evaluating site conditions.

[2] David Tomasko, Argonne National Laboratory, 9700 South Cass Avenue, Building 900, Argonne, IL 60439; Phone: (630) 252-6684; Fax: (630) 252-3611; e-mail: Dtomasko@anl.gov

Optimization of a monitoring system composed of 80 wells and 15 springs was performed following completion of remediation activities. A scoring system (0 to 10) for key sampling parameters was developed to provide an overall score (multiplicative product of all key parameters) for each well and spring to provide a quantitative estimate of the utility of the sampling location for post-remediation monitoring. Wells that had low overall scores were eliminated, leaving an optimized network for future sampling.

Key parameters for the groundwater wells included: (1) the location of the well relative to the location of the remediated source and the direction of groundwater flow; (2) the distance of the well from the remediated area; (3) the completion depth of the well relative to the vertical extent of contamination; and (4) historical trends in the contaminant concentration data. Wells that scored high were located a short distance down gradient of remediated source areas, were screened in the vertical zone of contamination, and had contaminant concentrations that exceeded applicable or relevant and appropriate (ARAR) guidelines.

Key parameters for springs included: (1) location or direct connection with contaminated source areas; and (2) historical trends in the concentrations for the contaminants of concern. Springs that scored high were directly linked with areas of former surface contamination and had concentrations that exceeded ARAR guidelines.

Plots of the overall scores for monitoring wells and springs were made, and average scores and standard deviations were calculated. A statistical threshold for retention in the monitoring system was set to the average score plus one standard deviation. Of the 80 initial wells and 15 springs, 6 wells and 7 springs were recommended for inclusion in the optimized monitoring system.

Using Geoscience and Geostatistics to Optimize Groundwater Monitoring Networks at the Savannah River Site

R. C. Tuckfield[3], E. P. Shine , R. A. Hiergesell , and M. E. Denham , Westinghouse Savannah River Co.
C. Beardsley, Bechtel Savannah River Inc.
S. Reboul, Westinghouse Savannah River Co.

A team of scientists, engineers, and statisticians was assembled to review the operation efficiency of groundwater monitoring networks at US Department of Energy Savannah River Site (SRS). Subsequent to a feasibility study, this team selected and conducted an analysis of the A/M area groundwater monitoring well network. The purpose was to optimize the number of groundwater wells requisite for monitoring the plumes of the principal constituent of concern, viz., trichloroethylene (TCE). The project gathered technical expertise from the Savannah River Technology

[3] R. Cary Tuckfield, PhD, Savannah River Technology Center, Westinghouse Savannah River Co., Bldg. 773-42A, Aiken, SC 29803; Phone: (803) 725-8215; e-mail: cary.tuckfield@srs.gov

Center (SRTC), the Environmental Restoration Division (ERD), and the Environmental Protection Department (EPD) of SRS.

The intent of this effort was to provide a technical basis report for the development of a RCRA part B permit modification for post closure care of the A/M area to be submitted to the South Carolina state regulator for approval. A multidisciplinary approach, combining geochemistry, geohydrology, geostatistics, and regulatory knowledge determined whether or not a well should remain on the current sampling schedule. The wells within each of three aquifer zones were evaluated with respect to relevancy, reliability, and regulatory importance. These evaluations identified sets of wells that were considered good candidates for deletion from the sampling schedule. The effects of less data due to well deletion were then evaluated using geostatistical redundancy analysis. In addition, historical trends in the contaminant concentration data were examined to determine those analytes that should remain on the sampling schedule for each well.

Major findings include:

- Of the 60 plume definition wells in the M-area (i.e. Water Table) aquifer, 22 are recommended for elimination from the sampling schedule.
- Of the 49 plume definition wells in the Upper Lost Lake aquifer, 9 are recommended for elimination from the sampling schedule.
- Of the 55 plume definition wells in the Lower Lost Lake aquifer, 11 are recommended for elimination from the sampling schedule.
- Eight analytes in 43 Point of Compliance (POC) and Background (BKGD) wells, 19 analytes in 199 Plume Definition (PD) wells, and an additional 4 analytes in 190 of the PD wells were recommended for elimination.

The purpose of this effort was to:

- assess the potential for reducing the number of wells within the selected network, number of chemical analytes required per groundwater sample, and the sampling frequency per well within the network without compromising the ability to accurately estimate the extent and direction of the contaminant plume and the efficacy of remedial efforts.
- provide a technically defensible basis for recommending changes, if any, to the SRS groundwater monitoring networks that will ensure greater efficiency without compromising the data necessary to support continued decision-making in the A/M Areas.

The A/M area monitoring well network has over 100 wells and analytes historically sampled under a RCRA Part B post closure care permit. It was regarded as having a substantial return on investment potential. The GME Team also reviewed the results from prior sampling reports to comprehensively assess the potential for improvement within this network.

Chemicals that were disposed of decades ago at the land surface in the A/M Area have since migrated downward into the saturated zone and are now being carried along with groundwater in the direction of the natural discharge zones. An extensive network of wells have been installed in the 1980's and 1990's to monitor the movement of contaminant plumes, primarily TCE and PCE, within this flow system.

TCE has spread further than the other contaminants and was therefore selected as the principal constituent of concern and contaminant for analysis.

Methods

The A/M Area network efficiency is assessed via a multidisciplinary approach, combining geochemistry, geohydrology, geostatistics, and regulatory knowledge. This approach is based on principles of relevancy, redundancy, and reliability, as well as regulatory requirements.

This approach is called the 4Rs Sampling Reduction Technology for monitoring well assessment. The 4Rs also provide a concise summary of the process used to define this technical basis document. The process is presented in phases but in practice it is iterative, using the information generated in one phase to reassess earlier results.

- Relevancy assessment uses geohydrological judgment and site-specific knowledge to identify wells that do not provide pertinent information about the contaminant plume.

- Reliability assessment investigates well performance based on the ability of a well to meet acceptable specification limits for pH and turbidity. It also evaluates any evidence for breeches in aquifer confining layers.

- Regulatory assessment means that wells automatically retained on the sampling schedule are point-of-compliance or background wells pertinent to RCRA regulations.

- Redundancy assessment (via geostatistics) identifies wells spatially located in close proximity to other monitoring wells. Such wells provide little additional monitoring information duplicated by nearby wells.

Analyte Reduction Analysis

In addition, the list of groundwater monitoring constituents contained in Appendix IIIB-A of the A/M Area HWMF Part B permit is inordinately lengthy. Isoconcentration maps were examined to determine whether or not areas of high contaminant concentration were present near any of the known source sites. In most cases, no such areas were identified indicating that the constituents in question had not been released to groundwater and did not require monitoring. In cases where this examination did reveal the presence of a contaminant plume, the size and location of the plume were used to determine which wells were appropriate to monitor the contaminant in question.

MAROS Decision Support Software for Long-Term Monitoring Optimization: Application To McClellan Air Force Base, California

Julia J. Aziz[4] and Charles J. Newell, Ph.D., P.E., Groundwater Services, Inc., Houston, TX

Hanadi S. Rifai, Ph.D., P.E and Meng Ling, University of Houston, Houston, TX

Jim Gonzales, Air Force Center for Environmental Excellence, Brooks AFB, TX

The Monitoring and Remediation Optimization System (MAROS) software provides non-statisticians with an easy-to-use tool for making long-term groundwater monitoring programs more efficient and effective. MAROS is a decision support tool based on statistical methods applied to site-specific data that account for relevant current and historical site data as well as hydrogeologic factors (e.g. groundwater velocity) and the location of potential receptors (e.g., wells, discharge points, or compliance boundaries). Based on this site-specific information the software helps identify monitoring wells that are i) sampled too frequently, or ii) are redundant because the well provides little additional information about the plume. The software is intended to serve as a combined data management/data analysis system for those involved in monitoring groundwater plumes that are not statistical experts.

At the McClellan site, the MAROS software applied primary lines of evidence (parametric and nonparametric trend analysis) to assess trends in historical site analytical data at the plume of interest. These lines of evidence were then consolidated and weighed to obtain synthesized site plume stability information, which was used for assessing future sampling duration, location and density to assist in identifying future compliance monitoring goals for McClellan AFB. More rigorous statistical sampling optimization methods (i.e. Delaunay Triangulation and Cost Effective Sampling) in the MAROS software were also applied to the McClellan site to assess the minimum number of wells, sampling frequency, and well density suggested for future compliance monitoring at the site. These preliminary monitoring optimization results provide a basis for which to make more cost effective, scientifically based future long-term monitoring decisions.

As the monitoring program at McClellan AFB proceeds, more recent sampling results can be added to historical data to assess the progress of the current monitoring strategy. The optimization process can be reviewed and updated periodically using the MAROS guidance recommendations. MAROS addresses a variety of groundwater contaminant plumes (e.g., fuels, solvents, metals) and is designed to be "evergreen" so that long-term monitoring plans can be modified as the plume changes over time (e.g., reducing monitoring efforts when a plume changes from stable to shrinking).

---

[4]Julia Aziz or Charles Newell, Groundwater Services, Inc., 2211 Norfolk St., Suite 1000, Houston, TX 77098-4044; Phone: (713) 522-6300; Fax: (713) 522-8010 web: www.gsi-net.com e-mail: jjaziz@gsi-net.com e-mail: cjnewell@gsi-net.com

Redundancy Analysis for 2000 GSAP at NAS Fort Worth, JRB

Theodore P. Lillys,[5] Patrick A. Sullivan, and Lynn Morgan, HydroGeoLogic, Inc.

Remedial and long term monitoring activities at Naval Air Station Fort Worth, Joint Reserve Base, Fort Worth, TX, have been focused on curtailing the migration of chlorinated solvent plumes in the groundwater, primarily TCE, beyond the site's eastern boundary. As part of the Year 2000 Groundwater Sampling and Analysis Plan (GSAP), HydroGeoLogic, Inc. proposed to the client, the Air Force Center for Environmental Excellence (AFCEE), to optimize the Year 2000 sampling plan based upon the application of advanced statistical techniques. Historically, over 260 sampling locations had been monitored on a quarterly basis for over 5 years at costs of $300-$400K per year. The objective was to minimize costs associated with sampling by eliminating sampling locations that do not contribute to the characterization of the TCE plume boundary, especially along the eastern boundary (see Figure 5.2 for this region of interest).

In groundwater monitoring the primary costs are attributed to sample acquisition, sample analysis, data management, and reporting. While minimizing these activities are an important component in reducing the overall cost of monitoring, minimizing the number of samples has, by far, the most significant impact on cost reduction, particularly where sites are over-sampled. In several cases (Lawrence Livermore Nation Laboratory, Savannah River, NAS Brunswick), reported annual cost reductions from 40% to 55% have been achieved by optimizing when and where to take groundwater samples.

To address this problem, HydroGeoLogic, Inc., has developed and implemented a state of the art application (Long-Term Monitoring Optimization (LTMO) ToolKit) that determines optimal sampling locations and frequency to meet site-specific monitoring goals and DQOs. The LTMO ToolKit utilizes geostatistical methods and temporal trending techniques to provide an objective and quantitative basis for selecting and justifying sampling plans that can reduce the number of required samples. In addition, the methodology can also address the issue of under-sampling by identifying locations where DQO's are not met. What follows is a brief description of the application of the LTMO ToolKit to optimize the year 2000 sampling plan at NAS Fort Worth, JRB, Fort Worth, TX.

Geostatistical methods (i.e., kriging) contained in the LTMO ToolKit were applied in the design of a new sampling strategy that minimizes the number of samples taken while meeting the data quality objectives. The methodology is comprised of three parts: data analysis, spatial estimation, and optimization. The objective of data analysis was to identify the underlying spatial correlation model. Spatial estimates are generated based upon the correlation model. The optimization analysis identifies redundant samples within sampling events. This methodology was successfully applied at NAS Fort Worth JRB in the development of the 2000

[5]Theodore P. Lillys, HydroGeoLogic, Inc, 1155 Herndon Parkway, Suite 900, Herndon, VA 20170; Phone: (703) 478-5186; e-mail: tpl@hgl.com

Groundwater Sampling and Analysis Program (GSAP). The proposed GSAP requires a total of 72 sampling locations as compared to 193 taken in 1999 (see locations in Figure 5.3), a reduction of over 60%.

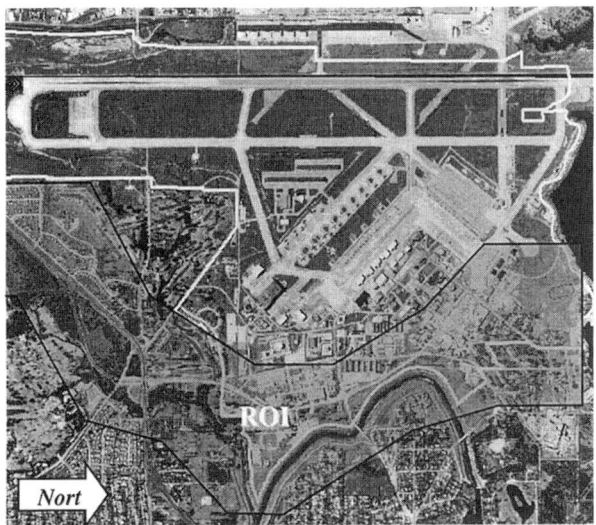

Figure 5.2. Site and region of Interest (ROI).

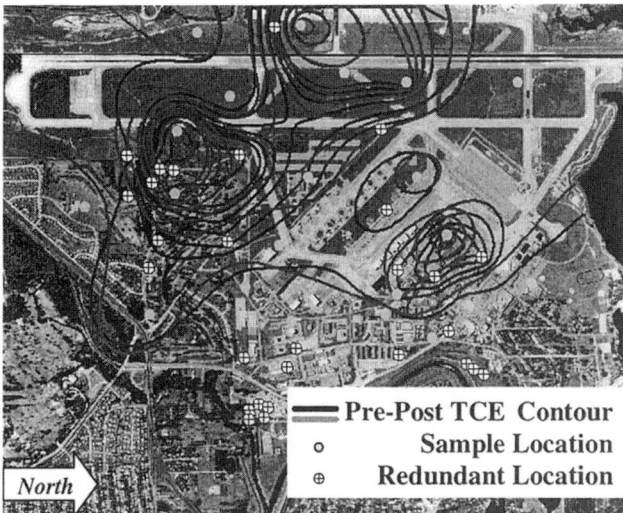

Figure 5.3. TCE Contours and sample locations before and after optimization.

Optimizing Long-Term Groundwater Monitoring Networks Using the Geostatistical Temporal/Spatial (GTS) Algorithm at the Massachusetts Military Reserve (MMR)

Dr. Kirk Cameron[6], MacStat Consulting, Ltd
Philip Hunter, P.G., Air Force Center for Environmental Excellence (AFCEE)

A spatial and temporal algorithm was developed for optimizing long-term monitoring (LTM) networks at U.S. Air Force installations. The primary objective was to determine the degree to which sampling, laboratory analysis, and/or well construction resources could be pared without losing key statistical information concerning the plumes being monitored.

Two plumes located at the Massachusetts Military Reservation (MMR) were used as a pilot project to develop the algorithm. Key contaminants of concern included EDB, Benzene, TCE, and PCE. Concentrations of these contaminants exceeded applicable maximum contaminant levels only a small fraction of the time, and generally were non-detect. Also, the "hits" were concentrated in a fairly small subset of the known monitoring wells. Consequently, the algorithm had to be able to handle highly skewed, uncertain, and spatially clustered concentration measurements.

To optimize an LTM network, an accurate assessment of ground-water quality over time is needed to construct an interpolated map of the concentration levels, and to accurately assess trends or other changes in individual monitoring wells. Changes in such maps over time indicate whether ground-water quality has improved or declined.

The optimization algorithm incorporates a decision pathway analysis that separately identifies temporal redundancy and spatial redundancy. Temporal redundancy (i.e., autocorrelation between closely spaced measurements) may be reduced by lengthening the time between sample collections. Spatial redundancy, indicative that too many wells are being monitored, may be reduced by removing selected wells from the network as long as the ability to assess ground-water quality is not sacrificed.

Part of the temporal algorithm involves computation of a composite temporal variogram to determine the least redundant overall sampling interval. Using this one-dimensional measure of autocorrelation between sampling events, the lag time at which the variogram reaches the sill is the sampling interval at which the same-well measurements become essentially uncorrelated and, therefore, non-redundant.

The spatial algorithm is predicated on the notion that well locations are redundant if nearby wells offer nearly the same statistical information about the underlying plume. A well was considered redundant if its removal did not significantly change: 1) an interpolated map of the plume, 2) the local kriging variances in that section of the plume, and 3) the average global kriging variance. To identify well redundancy, local kriging weights were accumulated into global weights and used to gauge each well's relative contribution to the interpolated plume map. By

---

[6]Dr. Kirk Cameron, MacStat Consulting, Ltd, 10330 Mill Creek Ct., Colorado Springs, CO 80908; Phone: (719) 532-0453 e-mail: kcmacstat@aol.com

temporarily removing that subset of wells with the lowest global kriging weights and re-mapping the plume, it was possible to determine how many wells could be removed without losing critical information.

At the test sites, close to 20 percent of the monitoring well locations were identified as spatially redundant. Furthermore, it was demonstrated that sampling frequencies could be relaxed from quarterly to annually and in some cases extended to once every five quarters. The overall reduction in the total annual sampling and analytical budget for these plumes was estimated to be between 35 and 40 percent or about $200,000 per year.

Long-Term Groundwater Monitoring Optimization Using a Time-Varying Bayesian Filter

Donna M. Rizzo[7] and David E. Dougherty[8], Subterranean Research, Inc.

Subterranean Research, Inc. (SRI) applied long term monitoring optimization (LTMO) at the Army SSCOM site in Natick, Massachusetts. This work was performed over a period of a few weeks to demonstrate new technologies. Due to the rapid pace of our project and the need for additional site characterization and related data, these results must be considered preliminary; follow-on analyses are planned and will incorporate the additional data. HydroGeoLogic, Inc. (HGL) gave SRI access to their calibrated MODFLOW-SURFACT 3-D simulation model for groundwater flow and solute transport, provided a conceptual understanding of the site and monitoring data from a number of existing monitoring wells, and collaborated with SRI to develop the preliminary LTMO objectives.

The time-varying Bayesian filter approach enables dynamic calibration of existing site simulation models to fit most recently collected data and to allow dynamic updates of the uncertainties as monitoring data are collected. Then, based on a current understanding of a site, a "model forecast" of concentrations and uncertainties is calculated to the time when new measurements are planned. The method has the advantage of directly incorporating measurement data as well as the spatial and temporal effects. Field measurements are used to adjust the model forecast by an amount proportional to the difference between actual field measurements and the model forecast of measurements. Both uncertainties and variances are adjusted based on measurement values and locations. The measurements can be obtained at irregular times and locations. See Rizzo et al. [2000] for more information.

A predominantly trichloroethene (TCE) plume extends on-and off-site and needs to be monitored. Monitoring data from a number of existing monitoring wells (MWs) exist (see Figure 5.4(a)) as well as a calibrated 3-D simulation model for groundwater flow and transport. The decisions that will be made are which MWs will

---

[7] Donna M. Rizzo, Ph.D. Subterranean Research, Inc., 372 Maple St., P.O. Box 1121, Burlington, VT, 05402-1121; Phone: (802) 658-8878; e-mail: drizzo@subterra.com

[8] David E. Dougherty, Ph.D., Subterranean Research, Inc., 33 Enterprise Place, Suite 5, Duxbury, MA 02331; Phone: (781) 934-7199; e-mail: **ddougher@subterra.com**

no longer be used and where would a replacement MW would best be located. The purpose of the MWs is to estimate concentrations at a Region of Interest (ROI), which in this case is a compliance boundary, formed by the northwest portion of the property line (Figure 5.4(a)). The selected site goals were to 1) predict TCE concentrations (specifically, estimate the 5 ppb contour near the ROI with 95% certainty), 2) find existing monitoring wells that could be by-passed during the current sampling schedule and not materially increase the uncertainties in concentration estimates at the ROI and 2) locate the best location for a replacement monitoring well.

Monitoring well data collected over a 6 year sampling history in which a large number of wells had observations were used in SRI's Bayesian Filter to develop estimates of concentrations and variances of estimation errors. The later were subjected to geostatistical redundancy analysis. Since the primary goal was to reduce the number of monitoring wells as much as possible without increasing the uncertainty at the ROI, total variance along the ROI was obtained by summing over 16 sentinel points.

The calculations were repeated, assuming that one of the original 15 monitoring wells was not used, to determine the increase in uncertainty for the ROI resulting from the omission of that one MW. Doing this for each monitoring well in turn, wells are identified that can be omitted from sampling and yet would not change the uncertainty associated with the predicted TCE concentrations for that particular sampling time. Results are collected in the bar plot of Figure 5.4(b). Projecting the results of this spatial redundancy analysis along with the results of the temporal variogram (not shown) suggest that 6 MWs can be removed from the monitoring schedule and that other MWs could be measured on a 1 to 2 year cycle, rather than quarterly. Figure 5.5 shows where the estimated 5 ppb contour is uncertain and where an additional MW can reduce uncertainty and maximize the value of data. This approach to risk visualization and planning allows for longer perspectives than the traditional diagnostic ("snapshot in time") geostatistical methods.

Acknowledgments

We are grateful to Steve Young at HydroGeoLogic, Inc. for fostering a productive collaboration and John McHugh at the Army SSCOM for his support of this LTMO project.

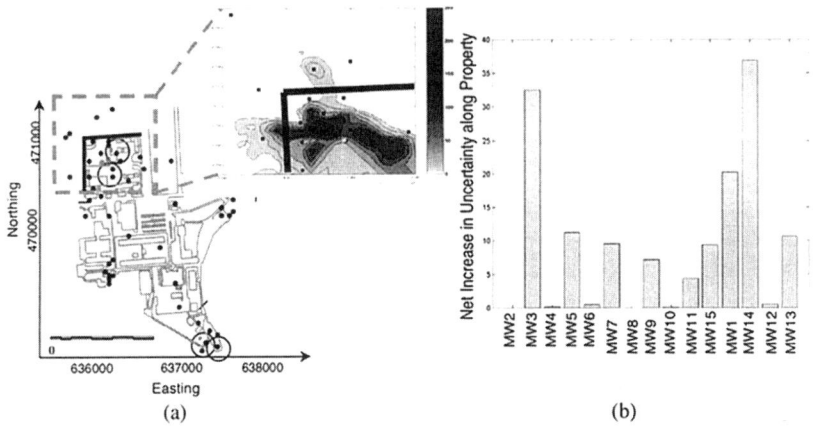

(a) (b)

Figure 5.4. (a) Site map and initial predicted TCE plume (maximum over all layers in model) and (b) bar plot of redundancy analysis indicating the 6 wells that may be removed from the current monitoring schedule.

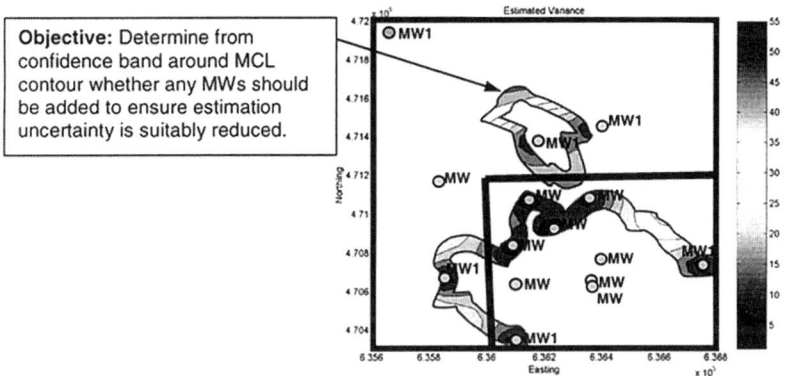

Figure 5.5. Confidence band around 5ppb TCE contour indicating where uncertainty is greatest.

73

Space-Time Optimization of a Long-Term Monitoring Network Using a Transport Model and a Kalman Filter

Graciela S. Herrera[9], Joseph Guaarnaccia, and George F. Pinder[10]

The optimization of an existing groundwater quality sampling network was obtained for a superfund site that formerly was a chemical plant. In Figure 5.6 the southeast portion of the plant is shown. As can be observed, a river is located close to the site and at some locations defines the limit of the property. The objective of the design was to optimize the existing sampling network while obtaining a contaminant concentration estimate with small error uncertainty in selected areas. The algorithm that we have developed (Herrera, 1998 and Herrera et al. 2000), that is based on the combined use of a stochastic transport model and a Kalman filter, can optimize sampling networks through three different strategies: the reduction of the number of sampling wells, sampling the same set of wells fewer times, or a combination of these two strategies. In essence, the method eliminates sampling positions and sampling times that are redundant. The sampling times can be irregular for each well that remains active in the network.

A contaminant plume that in some areas extends outside the property limits has been intensively monitored for more than fifteen years. Of the suite of chemicals present in the plume, chlorobenzene is the most ubiquitous. For this reason the design of the optimization of the sampling network was based on this chemical. The sampling program was designed for a period of 23 months. In the original sampling program a total of 322 samples from 62 sampling wells were considered. A maximum sampling frequency of 15 days is considered for the new sampling program, which gives a total of 46 possible sampling times for the 23 months.

The application was based on the use of a preexisting chlorobenzene transport model. The plume predicted for the model at the end of the period analyzed is shown in Figure 5.6. From the stochastic transport model a concentration estimate and its error variance are obtained. This variance is a measure of the uncertainty of the estimate obtained from the model. The objective of the design was to reduce in 96% the sum of variances of the concentration estimate errors at the end of the 23rd month for the risk zones. These two risk zones are located at regions where the site neighbors the river or populated areas, and they are shown with a bold black contour in Figure 5.6.

[9]Graciela Herrera, Mexican Institute for Water Technology, Paseo Cuahunáhuac 8532, Progreso, Jiutepec, Morelos 62550, MexicoPhone: (5-777) 3 19 4000 ext 805email: gherrera@tlaloc.imta.mx

[10]George F. Pinder, Research Center for Groundwater Remediation Design, College of Engineering and Mathematics, University of Vermont, Burlington, VT, 05405; e-mail: pinder@emba.uvm.edu

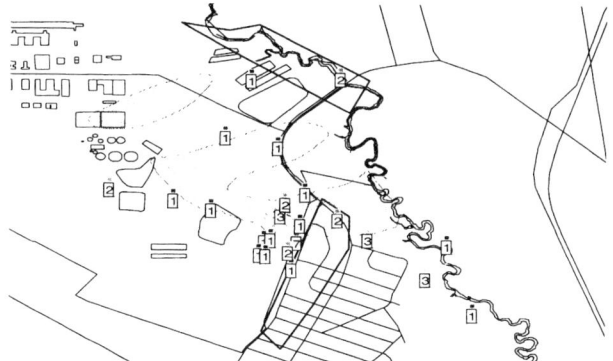

Figure 5.6. Study area, wells selected for the sampling network and number of sampling times included in the schedule of each well. The zones at which the total variance of the estimate error is minimized are shown with black contours and the chlorobenzene plumes obtained from the deterministic model at the end of the period of interest are shown with contours of lighter colors.

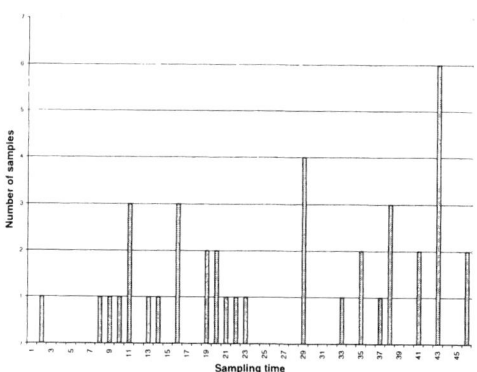

Figure 5.7. Number of samples selected at each possible sampling time. The bar height indicates the number of samples to be taken at the time indicated under its base. A total of 40 samples were required to reduce the error estimate variance in 95%.

From the 62 existing sampling wells, 23 were selected by the method proposed in this work. A total of forty samples, from the 322 samples included in the original sampling plan, were enough to reach the 96% reduction in the sum of error variances. The wells selected to remain active in the network are shown in Figure 5.6 with a square, the number shown close to each well is the total number of water samples that would be taken from the corresponding well during the 23 months period. Figure 5.7 shows the number of samples per sampling interval.

Chapter 6: Future Needs For Long-Term Monitoring Design

Improving the state of the art of long-term monitoring (LTM) and its design requires continued research and new implementation guidance to bring state-of-the-art methods into practice. This chapter presents recommendations in the areas of research and technology transfer for LTM design.

6.1 Research Needs

6.1.1. Develop methods for the integration and use of nontraditional data (e.g., sensor or screening technologies). Methods are required that can combine information with differing data types and degrees of uncertainty into an integrated data set that can be used for the design and operation of LTM systems. Traditional LTM programs require physical samples drawn from wells, which can involve substantial labor and materials costs. Alternative forms of data collection are being developed, such as *in situ* and wireline environmental sensors. Many of these sampling methods have lower accuracy or precision than traditional sampling procedures, but because of lower unit prices may be more spatially or temporally dense. Other methods are being developed that measure indicators (i.e., parameters other than the parameters of interest) that are then used as indicators for the actual parameter required. A dynamic field sampling approach advocates that such data be used with traditional samples (Crumbling et al., 2001; DOE, 2001) to produce more informative data sets for achieving overall decision quality objectives (i.e., producing data that is of sufficient quality to allow appropriate decisions to be made, taking into consideration both analytical accuracy and spatial and temporal representativeness). Disparate data types can be combined, too. For example, surface geophysical data sets offer dense spatial coverage but have significant uncertainties associated with measuring subsurface values. On the other hand, water samples analyzed at a laboratory have much greater accuracy but spatially are very sparse. The temporal or spatial scales may also differ substantially. The integration of these two data sets into a single set that provides the correct and appropriate information to designers and decision makers is difficult. To include these innovations in monitoring, methods that can integrate data from a variety of sources, accuracies, and types are needed.

6.1.2. Determine and characterize the variability in long-term monitoring data. The variability of laboratory analysis results, which can be caused by differences in methods, laboratories, instruments, and analysts, is both well documented and incorporated into environmental databases. Different field sampling methods can also have large variability in results. For example, contaminant concentrations in water samples obtained using micro-purge devices will differ from those in samples obtained from the same well using conventional "purge and sample" procedures. The length of a well's screened interval and its orientation (vertical vs horizontal wells) can also affect measured concentrations. Similar variability will be exhibited with innovative monitoring methods, as indicated above in 6.1.1, and the sources of variability will be technology-specific. Over long time horizons, sampling techniques and methods must be expected to change. When sampling methods change at a site, the historical records may show apparent anomalies, containing either unexplained abrupt increases or deceases in contaminant concentrations when sampling methods change. Research is needed to develop methods to understand, quantify, and establish policies for addressing these changes before they occur. For example, researchers might use both the current and replacement sampling methods during a specified time period in order to determine correlations between the methods and to identify parameters and methods for adjusting historical values to compare with results obtained using the new method. This type of research may result in procedures that require using dual sampling methods for a set period (or number of samples) when a new sampling method or procedure is introduced.

Similar research is needed to address other problems that are expected to occur as wells and other monitoring devices are observed over long time periods. Screen clogging, the development of bio-films in wells, changes in the near-well media, and the potential for foreign objects or substances in the wells are just a few examples of well-known long-term changes that affect the observed parameter values obtained from wells. Factors that affect the life-cycle accuracy, calibration, and errors of innovative in situ sensors need to be determined. In addition, contaminant concentrations at individual monitoring wells may exhibit short-term, weather-induced fluctuations superimposed on long-term trends, making identification of anomalies that can influence the observed long-term trend more difficult. For example, in a shallow unconfined aquifer, a recharge event can dilute contaminant concentrations, which rise again shortly thereafter. To address and analyze long-term trends, methods for identifying, evaluating, and responding to changes in the data that are not related to contaminant transport need to be developed.

6.1.3. Develop improved decision rules and protocols for long-term management. The methods presented in Chapter 4 illustrate that the current state of the practice in monitoring design focuses primarily on location and frequency of sampling a single constituent. Effective long-term management of sites requires more comprehensive decision rules and protocols that address (1) which constituents and indicator parameters should be measured and at what locations and frequencies; (2) what data collection, validation, and analysis procedures should be followed to collect that data; (3) what procedures to be followed when the data collected indicate a problem with the system being monitored; and (4) when the underlying conceptual model and the monitoring system itself need to be reevaluated or changed.

6.1.4. Develop and use "living" performance assessment models. "Living" models are models that are updated, refined, and used as new information becomes available. Environmental restoration sites can be complex, dynamic systems, and LTM optimization methods need to be developed that can both effectively analyze current conditions at the site and predict and accommodate the effects of observed changes in those conditions over the long term. Living models should also include continual or periodic reevaluation of the assumptions and goals used in the initial designs of the LTM and remediation systems and should update and refine the designs when needed, in addition to updating the model as new data become available.

During the initial remediation design phase, site characterization is performed and a conceptual model is developed. At some sites, simulation models based on this conceptual model are developed to help predict risk and to identify remedies, using historical data for calibration and validation. These models are then used for LTM design, as described in Chapter 4 (Sections 4.3 and 4.4). Once the LTM design phase is complete, however, these models are usually not updated and validated as new data are collected during the LTM operation phase.

Moreover, most LTM analyses focus only on conditions observed before or during design and are usually defined by the data collected during a single "snapshot" in time or at a single well over time. Using only these limited historical data, it is difficult to predict site-wide performance over time, which is needed to assess whether the remediation is performing as expected or whether intervention is required in the long term. To address these needs, methods should be developed for performance assessment that incorporate living models updated as information becomes available. These methods should be low cost, accessible to practitioners, and as automated as possible.

The term *performance assessment* is used broadly here, encompassing both hazardous waste sites and long-term waste disposal sites. At hazardous waste sites, performance assessment typically involves risk assessment, in which long-term effects are predicted under a variety of fixed exposure scenarios, often over 30-year compliance periods. At waste disposal sites with long compliance periods (hundreds or thousands of years), performance assessment involves both risk assessment and failure analysis from hazards such as earthquakes, volcanoes, and 100-year floods, as well as from changes in exposure scenarios due to societal, political, and technological changes. Living performance assessment models are needed for both of these types of sites.

These recommendations are consistent with those of the National Research Council (NRC) Committee on Subsurface Contamination at DOE Complex Sites. The NRC committee, in a report on research needs in subsurface science within the U.S. Department of Energy's Environmental Management Science Program (NRC, 2000a), recommended that new methods be developed that can "detect both current conditions and changes in system behaviors. These methods may involve the application of conceptual, mathematical, and statistical models to determine the types and locations of observation systems and prediction of the spatial and temporal resolutions at which observations need to be made" (NRC, 2000a, p. 122). The NRC committee also recommended that model validation be studied, including identifying what data must be collected and how to use the data to determine whether a model is valid.

6.1.5. Develop improved methods for linking remediation processes and LTM. As highlighted in Chapter 5, LTM optimization is inextricably linked to the remediation process used, but this link has not been widely acknowledged in existing LTM design methods. The remediation technique used at a site, and its performance assessment requirements, will have a significant influence on the amount and type of monitoring required. Data collected for LTM may indicate that changes are needed in the system's operation or configuration. It should be apparent that the remediation goals provide the linkage between remedy and LTM selection, design, and operation. Procedures for incorporating LTM requirements into the remedial design process and for optimizing the LTM appropriately are needed.

There is also a need to better incorporate LTM data collection into the on-going operations. For example, flow containment is a remedial design objective at many sites and groundwater flow models are often used to generate the remediation system. The predicted flow field used for design is based on limited field data and/or flow modeling and will therefore have errors. These errors translate into differences between predicted and actual remedy performance. Better methods for assessing performance and improving operations management are needed. Hydraulic controls are almost always surrogates for contaminant transport control and risk (i.e., rarely are water levels the concern, but water levels are observed to ensure that contaminant transport is within design limits). Additional research is needed to demonstrate the effectiveness of water level measurements as surrogates to demonstrate contaminant transport control and to acknowledge the link to contaminant transport by updating the conceptual model as water levels (and flow field) dynamics are better defined through LTM.

6.1.6.   Develop methods for designing LTM systems at sites with fractured bedrock or karst terrain. Most of the spatial interpolation methods used in LTM optimization in Chapter 4 assume spatially continuous media (that is, locations that are spatially close have similar properties). These assumptions are not valid for fractured and dissolution bedrock sites. Temporal trend analyses, noted in Chapter 4, can be used at fractured sites, but additional methods are needed. Methods such as tracer tests structured specifically for fractured or karst terrain are routinely used for assessing spatial continuity or connectivity. Geostatistical methods that better incorporate linear and planar connectivity embedded in the three-dimensional subsurface are needed before performing redundancy studies, like those noted in Chapters 4 and 5, in fractured media.

6.1.7.   Improve methods for failure analysis and robust LTM design. Most current LTM design methods do not incorporate formal safety factors to acknowledge the uncertainties known to be in the system. Other engineering disciplines (even ones that deal with subsurface geologic uncertainty, such as geotechnical engineering) have formal systems for acknowledging these uncertainties in design procedures and for "overdesigning" systems to account for potential failure. This is especially important in "optimal" design, which develops a system with as few redundancies (or excess capacity) as allowed by the problem specification in order to reduce costs and maximize efficiency. This approach can result in systems that are designed as close to failure as possible, while still meeting design objectives. Methods for robust optimization, which identify designs that are predicted to perform well under a range of plausible scenarios, are available but are not widely used, primarily because of their complexity and large computational requirements. One way of incorporating these concepts is to use a safety factor. A safety factor is applied to design variables or constraints to provide a margin of safety to account for uncertainty. For example, in one method of structural design for buildings, the estimated loads are increased by a multiplier called the *safety factor* to account for uncertainties or to consider unmodeled design objectives. However, there is no direct analog for structural loads in the contaminant transport field. Previous research results have been negative and skeptical about the use of safety factors in the groundwater field, suggesting that the proper safety factor is site-specific and that stochastic analyses are needed to establish safety factors on a site-by-site basis. This issue deserves further study to determine its applicability to groundwater monitoring and remedial design and, if applicable, to develop standard procedures for incorporating these concepts.

6.1.8. Improve electronic management of and access to data. Before the methods suggested previously can gain widespread use, site data must be readily available for analysis in electronic format. Improved methods for web-based acquisition, screening, recording, and evaluating of monitoring data need to be developed and disseminated, including automatic screening and evaluation. New environmental data deliverable (EDD) templates should be developed in parallel with new sensors; for example, on-line recalibration of sensors needs to be appropriately annotated in the data set.

6.1.9. Create publicly available data archives and test data sets that can be used for LTM research, development, and education. Developing and testing new LTM design methods currently requires time-consuming data acquisition and cleaning. New analytical methods are rarely compared and benchmarked on the same data set as existing methods, which makes comparison and selection difficult. Publicly available data archives need to be created to address these difficulties. These archives need to be well tested, documented, and easily available for use by researchers and practitioners.

These data sets can be used to develop and test new methods or as tools for those who want to learn how to use the latest methods with a sample data set. A variety of data sets are needed with different complexity and temporal and spatial scales. As noted in Chapter 4, some LTM design methods are most appropriate for sites with substantial amounts of data whereas others are for sites with little data; data sets are needed for testing both types.. Realistic synthetic data sets, in which data are "known" or can be generated at all points in space and time, can also be helpful for validating new methods for LTM design.

The development of the data archives should include research on methods for data validation, typical error types, data storage, and data access. It is also important that the data sets not be "scrubbed"; that is, contradictory and anomalous data should not be removed from the archive unless the data are proven to be incorrect. Instead, multiple data set versions, representing the sequential results of quality assurance/checking procedures, should be present in the archive, together with documentation of the procedures employed to move from one version to the next.

Developing and maintaining such data archives will involve a number of challenges that need to be addressed. For research with data archives to be viable, archive storage locations must be developed along with procedures for storing, maintaining, and reviewing posted data. Major issues include allowable formats of the metadata and of the data itself; the amount of supporting information provided, such as site maps and reports; resources for documentation, distribution, and maintenance (server, staff, review procedures, etc.); reluctance to release remediation data because of social concerns (liability and litigation concerns, public relations concerns, etc.); and the ability (or inability) to easily publish "data" papers in peer-reviewed journals.

6.2 Technology Transfer Needs

The committee strongly encourages the development of uniform LTM guidance, regulation, and practice across regulatory jurisdictions. To achieve this goal, the following technology transfer developments are needed.

6.2.1. Regulatory guidance on the amount and types of data needed to demonstrate system integrity or performance. There is a perceived lack of regulatory guidance on the types of data analysis that are required to assess remediation performance and to identify appropriate long-term monitoring plans. At some sites, volumes of raw data are simply presented to the regulators as evidence for remediation performance. At other sites, substantial data analysis and modeling are performed. There is a need for guidance on the minimum level of analysis that should be performed to assess remediation performance during periodic site reviews.

6.2.2. Regulatory guidance and acceptance of dynamic sampling (or operation) plans. At some sites, regulators have approved dynamic sampling plans that change in response to new data. In other cases, fixed sampling plans were required and could only be changed with a formal modification. Dynamic work plans require that an accepted decision process be approved, rather than a fixed sampling (or operation) plan. For example, the EPA Technology Innovation Office has presented a new "Triad" approach that addresses the need for dynamic sampling plans (Crumbling et al., 2001). Regulatory agencies, responsible parties, and their consultants need to examine this issue in more detail and to develop guidance, examples, and procedures for using dynamic sampling approaches at sites.

6.2.3. Regulatory guidance on incorporating LTM needs early in the remedial process. Guidance should be developed on ways to incorporate LTM issues and needs early into the remedy selection and design process (or even into the initial characterization phase). At most sites, LTM optimization and remedy selection and design are performed separately but, as noted previously, the analyses are inextricably linked. There is a need for policy change to allow these two types of analyses to be considered simultaneously in an overall performance assessment.

6.2.4. Professional guidelines and education. Professional guidelines and education should be developed on the special problems associated with LTM and the methods available, using mechanisms such as workshops and regulator guidance. Many reviewers of this report requested that detailed step-by-step guidelines be developed on how to select and implement the methods described in Chapter 4; that was beyond the scope of this effort and is clearly a need for future work. This recommendation is similar to one made in the NRC's Committee on Intrinsic Remediation (NRC, 2000b), although their recommendation was specific to natural attenuation sites.

The guidelines and education should also address the following additional questions. (1) How much site characterization is needed prior to undertaking LTM design? (2) How can sensor and screening data be incorporated into LTM to meet decision quality objectives? (3) What criteria can be used to eliminate sampling for particular constituents at existing wells? (4) When can LTM be halted? (5) How should the public be involved in LTM optimization, particularly in defining LTM objectives? (6) What should be done with monitoring wells that are no longer being sampled—should they be maintained for potential future reuse or should they be sealed?

References

AFCEE. (1997). Final LTM optimization guide. U.S. Air Force Center for Environmental Excellence, San Antonio, Texas.

AFCEE. (1997). *Long-Term monitoring optimization guide, version 1.1*, Brooks Air Force Base, TX. http://www.afcee.brooks.af.mil/er/download/ltmfiles.exe).

Anderson, M. P., and Woessner, W. W. (1992). *Applied groundwater modeling: Simulation of flow and advective transport.* Academic Press, San Diego, CA.

Anné, D. C. (1992). "Know your detection and quantitation limits." *Pollution Engineering*, May 1, 68-71.

Anné, D. C. (1997). "Do you know the quality of your analytical data?" *Pollution Engineering,* http://www.pollutionengineering.com/archives/1997/pol0201. 97/02adp3f0.htm (February 1, 2001).

Arsenault, M., and M. Legg. (2001). "Electronic deliverable format, version 1.2I." www.arsenaultlegg.com/edds.htm, www.enabl.com, ALI.

ASCE Task Committee on Geostatistical Techniques. (1990a). "Review of geostatistics in geohydrology I: Basic concepts." *Journal of Hydraulic Engineering*, 116(5), 612-632.

ASCE Task Committee on Geostatistical Techniques. (1990b). "Review of geostatistics in geohydrology II: Applications." *Journal of Hydraulic Engineering*, 116(5), 633-658.

ASTM. (1996). *Provisional standard guide for developing appropriate statistical approaches for ground-water detection monitoring programs.* American Society for Testing and Materials, West Conshohocken, PA.

Aziz, J. J., Newell, C. J., Rifai, H. S., Ling, M., and Gonzales, J. R. (2000). *Monitoring and remediation optimization system (MAROS) software user's guide,* Air Force Center for Environmental Excellence, Brooks Air Force Base, TX.

Barry, D. A., and Sposito, G. (1990). "Three-dimensional statistical moment analysis of the Stanford/Waterloo Borden tracer data." *Water Resources Research*, 26(8), 1735-1747.

Bens, C., Mispagel, M., and Weisskopf, C. (1997). "Good laboratory practices (GLPs) for the analytical laboratory: QA/QC principles and practices." IAEA, El Salvador.

Bogardi, I., Bardossy, A., and Duckstein,, L. (1985). "Multicriterion network design using geostatistics," *Water Resources Research*, 21(2), 199-208.

Buller, R. D., Gradet, A., and Reed, V. S. (1984). "Ground water monitoring system design in a complex geologic setting." In *Proceedings of the Fourth National Symposium on Aquifer Restoration and Ground Water Monitoring*, National Water Well Association, Worthington, OH, 186-194.

Burrows, R. and Hall, J. (1997). "The method detection limit: Fact or fantasy." 13th Annual Waste Testing & Quality Assurance Symposium, July 6-9, 1997, Crystal City, Virginia, 200-203.

Buscheck, T. E., and Alcantar, C. M. (1995). "Regression techniques and analytical solutions to demonstrate intrinsic bioremediation." *Intrinsic Bioremediation*, ed. R. E. Hinchee, J. T. Wilson and D. Downey, Battelle Press, Columbus, Ohio, 109-116.

Cameron, K., (1999). "Optimization of LTM Networks: Statistical approaches to spatial and temporal redundancy," In U.S. Federal Remediation Technologies Roundtable.

Cameron, K., and Hunter, P. (2000). *Optimization of LTM networks using GTS: Statistical approaches to spatial and temporal redundancy*, Air Force Center for Environmental Excellence, Brooks AFB, TX. (Online at http://www.afcee. brooks.af.mil/er/rpo/GTSOptPaper.pdf).

Canales, T., and Ottesen, P. (1996). "Rapid data access: Key to integrated use of environmental characterization and monitoring information." Spectrum Conference 1996, http://www-erd.llnl.gov/library/spec/canales2.pdf.

Cave, M. R., Butler, O., Cook, J. M., Cresser, M. S., Garden, L. M., and Miles, D. L. (2000). "Environmental analysis." *Journal of Analytical Atomic Spectrometry*, 15(2), 181-235. Chou, C. J., O'Brien, R. F., and Barnett, D. B.

Christakos, G., and Killam, B. R. (1993). "Sampling design for classifying contaminant level using annealing search algorithms," *Water Resources Research*, 29(12), 4063-4076.

Cienawski, S. E., Eheart, J. W., and Ranjithan, S. (1995). "Using genetic algorithms to solve a multiobjective groundwater monitoring problem." *Water Resources Research*, 31(2), 399-409.

Clesceri, L. S., Eaton, A. D., and Greenberg, A. E. (1999). "Standard Methods for Examination of Water & Waste Water, 20th ed."

Cronce, R. C., Glass, S. A., and Trentacoste, N. P. (1999). "RAO/LTM Optimization at the Naval Industrial Reserve Ordnance Plant, Fridley, Minnesota—A case history." In USEPA, 2000, pp. 29-30.

Crumbling, D. M., Groenjes, C., Lesnik, B., Lynch, K., Shockley, J., Vanee, J., Howe, R., Keith, L., and McKenna, J. (2001). Managing uncertainty in environmental decisions, *Environmental Science and Technology,* 35(19), 404A-409A.

Davis, C. B., and McNichols, R. J. (1994a). Ground water monitoring statistics update: part I: Progress since 1988. *Ground Water* (fall), 148-158.

Davis, C. B., and McNichols, R. J. (1994b).Ground water monitoring statistics update: part II: Nonparametric prediction limits. *Ground Water* (fall), 159-175.

Davis, C. B. (1994). Environmental regulatory statistics. In *Handbook of Statistics,* Volume 12, Chapter 26, p. 817-865, Patil, G. P. and Rao, C. R., eds., Elsevier Science, New York.

de Marsily, G. (1986). *Quantitative hydrogeology,* Academic, San Diego, CA.

de Marsily, G. (1986). *Quantitative Hydrogeology.* Academic Press, Orlando, USA.

Delhomme, J. P., 1979. "Spatial variability and uncertainty in groundwater flow parameters: A geostatistical approach." *Water Resources Research,* 15 (2), 269-280.

Department of Energy Innovative Technology Summary Report DOE/EM-0592: Adaptive Sampling and Analysis Programs (ASAPs), OST/TMS ID 2946, Characterization, Monitoring, and Sensor Technology Crosscutting Program and Subsurface Contaminants Focus Area, August 2001. http://www.em.doe.gov/ost.

Deutsch, C. V., and Journel, A. G. (1998). *GSLIB geostatistical software library and user's guide,* 2nd ed., Oxford University Press.

Dishman, T. (1998). "Spreadsheet vs. databases: Which one offers the best view of your data." *Office Computing,* 9(8).

DOE. (2001). A report to Congress on long-term stewardship: Volume I—Summary report, Office of Environmental Management, Office of Long-Term Stewardship, United States Department of Energy, Washington, D.C., DOE/EM-0563, January.

Dougherty, D. E., and Marryott, R. A. (1991). "Optimal groundwater management: 1. Simulated annealing." *Water Resources Research,* 27 (10), 2493-2508.

Ely, D. M., Hill, M. C., Tiedeman, C. R., and O'Brien, G. M. (2000). "Evaluating observations in the context of predictions for the Death Valley regional groundwater system." *Proceedings of the 2000 Joint Conference on Water Resources Engineering and Water Resources Planning and Management*, Minneapolis, MN, compact disk, American Society of Civil Engineers, Washington, DC.

EPA. (1985). "EPA laboratory data validation functional guideline for evaluating pesticide/PCBs." EPA, Washington, D.C.

EPA. (1988). "EPA Laboratory data validation functional guideline for evaluating inorganics analyses." EPA, Washington, D.C.

EPA. (1991). "EPA Laboratory data validation functional guideline for evaluating organic data review." EPA, Washington, D.C.

EPA. (1993). Guide for evaluating technical impracticability of ground water restoration. OSWER Directive 9234.2-25, EPA-540-R-93-080, NTIS Order Number PB93-963507.

EPA. (1994). "Guidance for the data quality objective process." EPA/600/R-96/055, EPA, Washington, D.C.

EPA. (1996). Presumptive response strategy and ex-situ treatment technologies for contaminated ground water at CERCLA sites, final guidance. OSWER Publication 9283.1-12, EPA/540/R-96/023, NTIS Order Number PB96-963508.

EPA. (1996). A guide to preparing superfund proposed plans, records of decision, and other remedy selection decision documents. OSWER Memorandum 9200.1-23P, EPA 540-R-98-031, NTIS Order Number PB98-963241. (Online at http://www.epa.gov/superfund/resources/remedy/rods/index.htm)

EPA. (1997). "EPA Region 9 data quality oversight at the Aerojet Superfund Site: Executive summary." #6400044, EPA, Washington, D.C.

EPA. (1998). "Guidance for data quality assessment." EPA/600/R-96/084, EPA, Washington, DC.

EPA. (1998). "Field sampling and analysis technologies matrix, version 1.0." http://www.clu-in.org/pub1.htm.

EPA. (2000). "EPA quality manual for environmental programs." 5360 A1, http://www.epa.gov/quality/qs-docs/5360.pdf, EPA, Washington, DC.

Eppstein. M. J., and Dougherty, D. E. (1996). "Simultaneous estimation of transmissivity values and zonation," *Water Resources Research*, 32(11), 3321.

Eppstein, M. J., and Dougherty, D. E. (1998). "Optimal 3-D traveltime tomography," *Geophysics*, 63(3), 1053-1061.

Eppstein, M. J., and Dougherty, D. E. (1998a). "Efficient 3-D Data Inversion: Soil Characterization and Moisture Detection from Crosswell GPR at the ARA Vermont Test Site." *Water Resources Research*, 34(8), pp. 1898-1900.

Farrar, J. W., and Long., H. K. (1997). "Report on the U.S. Geological Survey's evaluation program for standard reference samples distributed in September 1996." 97-20.

Foote, K. E., and Huebner, D. J. (1996). "Managing error." University of Colorado at Boulder, Boulder, CO.

Forman, R. and Vitale, J. (1999). "Lessons learned from performance evaluation studies." 15th Annual Waste Testing & Quality Assurance Symposium, July 18-22, 1999, Crystal City, Virginia, 38-46.

Freeze, R. A., L. Smith, G. de Marsily, and J. W. Massmann (1988), "Some uncertainties about uncertainty," *Proceedings of AECL/-DOE '87 Conference on Geostatistical, Sensitivity, and Uncertainty Methods for Groundwater Flow and Radionuclide Transport Modeling*, San Francisco, CA, pp. 231-260.

Freeze, R. A., Massmann, J. Sperling, T., and James, B. (1990). "Hydrogeological decision analysis: 1. A framework," *Ground Water*, 28(5), 738-766.

Freeze, R. A., James, B. Massmann, J., Sperling, T., and Smith, L. (1992). "Hydrogeological decision analysis: 4, the concept of data worth and its use in the development of site investigation strategies," *Ground Water*, 30(4), 574-588.

Garbajosa, J., Alarcon, P. Garcia, H. Alandes, M., and Piattini, M. (2000) "Introducing the data role in models for database assessment." Product Focused Software Process Improvement, June 20-22, 2000, Oulu, Finland, 49-58.

Garner, W. Y., Barge, M. S., and Ussary., J. P. (1992). "Good laboratory practice standards; applications for field and laboratory studies." American Chemical Society, Washington, DC.

Gibbons, R. D. (1994). *Statistical methods for groundwater monitoring.* Wiley, New York.

Gilbert, R. O. (1987). *Statistical methods for environmental pollution monitoring.* Van Nostrand Reinhold, New York.

Goldberg, D. E. (1989). *Genetic algorithms in search, optimization, and machine learning.* Addison-Wesley-Longman-Reading, MA.

Graham, W., and McLaughlin, D. (1989). "Stochastic analysis of nonstationary subsurface solute transport, 2, Conditional moments," *Water Resources Research*, 25(11), 2331-2355.

Greenwald, R. (1999). "Hydraulic optimization demonstration for groundwater pump-and-treat systems." Washington, DC : U.S. Environmental Protection Agency, Office of Research and Development, Office of Solid Waste and Emergency Response.

Hansen, J. P. (1999). "How the Badger Army Ammunition Plant saved $400,000 in long-term monitoring costs." In USEPA, 2000, p. 61.

Hassig, N., Vail, Bates, L. D., Brown, M. Moke, C., and Michael, D. (1999). "Using the data quality objective process to revise a groundwater monitoring program: The experience at Pantex." In USEPA, 2000, pp. 53-54.

Herrera, G. S., and Pinder, G. F. (1998). "Cost-effective groundwater quality sampling network design," *Proc. XII International Conference on Computational Methods in Water Resources, Vol. I*, Crete, June 1998.

Herrera, G. S. (1998). *Cost effective groundwater quality sampling network design*. PhD.Thesis, University of Vermont, Burlington, Vermont.

Herrera, G. (1998). "Cost effective groundwater quality sampling network design," PhD Dissertation, Department of Mathematics and Statistics, University of Vermont, Burlington, Vermont, 1998.

Herrera, Graciela, Joseph Guarnaccia and George F. Pinder, 2000. A methodology for the design of space-time groundwater quality sampling networks, *Proceedings of the Conference Computational Methods in Water Resources XIII, vol. 1, Computational methods, subsurface and transport*, Eds. L. R. Bentley, et al., Balkema, Rotterdam, 579-585.

Hill, M. C., Ely, M. D., Tiedeman, C. R., D'Agnese, F. A., Faunt, C. C., and O'Brien, B. A. (2000). "Preliminary evaluation of the importance of existing hydraulic-head observation locations to advective-transport predictions, Death Valley regional flow system, California and Nevada." U.S. Geological Survey Water-Resources Investigations Report 00-4282, http://water.usgs.gov/pubs/wri/wri004282/.

Hillier, F. S., and Lieberman, G. J. (1995). *Introduction to operations research.* McGraw-Hill, New York.

Holland, J. H. (1975). *Adaptation in natural and artificial systems.* University of Michigan Press, Ann Arbor, MI.

Hollander, M., and Wolfe, D. A. (1973). *Nonparametric statistical methods.* Wiley, New York.

Hsueh, Y. W., and Rajagopal, R. (1988). "Modeling ground-water quality decisions." *Ground-Water Monitoring Review,* 8(4), 121-134.

Hudak, P. F., and Loáiciga, H. A. (1993). "An optimization method for monitoring network design in multilayered ground-water flow systems." *Water Resources Research,* 29(8), 2835-2845.

Hudak, P. F., Loáiciga, H. A., and Marino, M. A. (1995). "Regional-scale ground-water quality monitoring via integer programming." *Journal of Hydrology,* 164, 153-170.

Hudak, P.F. (1998). A method for designing configurations of nested monitoring wells near landfills. *Hydrogeology Journal,* 6, 341-348.

Hudak, P. F. (1999). "A method for designing up-gradient groundwater monitoring networks." *Environmental Monitoring and Assessment,* 57, 149-155.

Hunter, P. (1999). "Long-Term monitoring and optimization of an air force pump and treat facility." In USEPA, 2000, pp. 32-33.

Ismail, W. M. (2000). "GIS Database Design and Organization," Universiti Sains Malaysia, Malaysia.

James, B. R., and Freeze, R. A. (1993). "The worth of data in predicting aquitard continuity in hydrogeological design," *Water Resources Research,* 29(7), 2049-2065.

James, B. R., and Gorelick, S. M. (1994). "When enough is enough: The worth of monitoring data in aquifer remediation design," *Water Resources Research,* 30(12), 3499-3513.

Jenkins, R. A., Bayne, C. K., Maskarinec, M. P., Johnson, L. H., Holladay, S. K., and Patten, B. A. (1998). "Database vs. spreadsheet is no contest." *Phoenix Business Journal,* October 19, 1998.

Johnson, N. L. and Leone, F. C. (1977). *Statistics and experimental design in engineering and the physical sciences.* Vol. I. Wiley, New York.

Johnson, V. M., Ridley, M. N. Tuckfield, R. C., and Anderson, R. A. (1996). "Reducing the sampling frequency of ground water monitoring wells," *Environmental Science and Technology,* 30, 355-358.

Klawitter, E., and Barry, M. (1999). "Long-Term monitoring optimization at Naval Air Station, Brunswick, Maine." In USEPA, 2000, pp. 34-35.

Kalman, R. E. (1960). "A new approach to linear filtering and prediction problems." *Trans. ASME, Journal of Basic Engineering*, pp. 35-45.

Kendall, M. G. (1975). *Rank correlation methods*, 4th ed., Griffin, London.

Kendall, M. (1977). *The advanced theory of statistics,* Vols. I & II, 4th ed., Griffin, London.

Kirkpatrick, S., Gelatt, C. D., and Vecchi, M. P. (1983). "Optimization by simulated annealing," *Science*, 220, 617-680.

Loaiciga, H. A., Charbeneau, R. J., Everett, L. G., Fogg, G. E., Hobbs, B. F., and Rouhani, S. (1992). "Review of ground-water quality monitoring network design," *Journal of Hydraulic Engineering*, 118(1), 11-37.

Loáiciga, H. A. (1989). "An optimization approach for ground-water quality monitoring network design." *Water Resources Research*, 25(8), 1771-1780.

Lucas, J. M. (1982). "Combined Shewhart-CUSUM quality control schemes". *Journal of Quality Technology,* 14, 51-59.

Mahar, P. S., and Datta, B. (1997). "Optimal monitoring network and ground-water-pollution source identification." *Journal of Water Resources Planning and Management*, 123(4), 199-207.

Marin, C. M., Medina, Jr., M. A., and Butcher, J. B. (1989). "Monte Carlo analysis and Bayesian decision theory for assessing the effects of waste sites on groundwater, I, theory," *Journal of Contaminant Hydrology*, 5, 1-13.

Marryott, R. A., Dougherty, D. E., and Stollar, R. L. (1993). "Optimal groundwater management: 2. Application of simulated annealing to a field-scale contamination site." *Water Resources Research*, 29, (4) 847-860.

Massmann, J., and Freeze, R. A. (1987). "Ground-water contamination from waste-management sites: The interaction between risk-based engineering design and regulatory policy, 1. Methodology." *Water Resources Research*, 23(2), 351-367.

McDonald, M. G., and Harbaugh, A. W. (1988). "A modular three-dimensional finite-difference ground-water flow model." Techniques of water resources investigations of the United States Geological Survey, book 6, chapter A1, U.S. Geological Survey, Washington, D.C.

McDonald, M.G., and Harbaugh, A.W., 1988. *A Modular Three-Dimensional Finite-Difference Ground-Water Flow Model.* U.S. Government Printing Office, Washington, D.C.

McGrath, W. A., Dougherty, D. E., Pinder, G. F., and Rizzo, D. M. (1996). *FOCuS user's guide, version 1.0*, Research Center for Groundwater Remediation Design, University of Vermont, Burlington, VT.

Metropolis, N., Rosenbluth, A. W., Rosenbluth, M. N., Teller, A. H., and Teller, E. (1953). "Equations of state calculations by fast computing machines." *Journal of Chemistry and Physics*, 21(6), 1087-1091.

Meyer, P. D., and Brill, E. D. (1988). "Method for locating wells in a ground-water monitoring network under conditions of uncertainty." *Water Resources Research*, 24(8), 1277-1282.

Meyer, P. D., Valocchi, A. J., and Eheart, J. W. (1994). "Monitoring network design to provide initial detection of groundwater contamination." *Water Resources Research*, 30(9), 2647-2659.

Miller, R. G. Jr. (1981). *Simultaneous statistical inference*, 2nd edition, Springer-Verlag, New York.

Merkhofer, M. W., and Runchal, A. K. (1988). "Probability encoding: Quantifying judgmental uncertainty over hydrologic parameters for basalt," *Proceedings AECL/DOE '87 Conference on Geostatistics, Sensitivity and Uncertainty Methods for Groundwater Flow and Radionuclide Transport Modeling*, San Francisco, CA, pp. 629-648.

Naber, S., Buxton, B., McMillan, N., and Soares, A. (1997). "Statistical method for assessing the effectiveness of intrinsic remediation." *Proceedings of the Fourth International Symposium on In Situ and On-Site Bioremediation*, 5, 348-354, Battelle Press, Columbus, OH.

Naber, S. J., Buxton, B. E., and Herberholt, A. M. (1999). "Optimizing a Ground-Water Monitoring Network for Assessing Air Sparging Effectiveness on BTEX and MTBE." In USEPA, 2000, pp. 55-56.

National Research Council. (1993). In situ bioremediation: When does it work? National Academy Press, Washington, D.C.

National Research Council. (2000a). *Research Needs in Subsurface Science: U. S. Department of Energy's Environmental Management Science Program*, National Academy Press, Washington, DC. http://www.nap.edu.

National Research Council. (2000b). Natural Attenuation for Groundwater Remediation, National Academy Press, Washington, DC. http://www.nap.edu.

Olea, R. (1984). "Sampling design optimization for spatial functions." *Mathematical Geology*, 16(4), 365-391.

Quinlan, J. F. (1990). "Special problems of ground-water monitoring in karst terrains." In *Ground Water and Vadose Zone Monitoring ASTM STP 1053*, Nielsen, D.M. and Johnson, A.I., eds., American Society for Testing and Materials, Philadelphia, 275-304.

Patten, B. (1998). "The Great Spreadsheet/database debate continues." *Washington Business Journal*, November 2, 1998. http://www.els.salford.ac.uk/geog/staff/nmt/fungis/hsd/md3sect6.pdf.

Reed, P. M. (1999). "Cost effective long-term groundwater monitoring design using a genetic algorithm and global mass interpolation," M. S. Thesis, Department of Civil and Environmental Engineering, University of Illinois, Urbana, IL.

Reed, P. M., Minsker, B. S., and Valocchi, A. (1999). "Cost-effective long-term monitoring design for intrinsic bioremediation," *Proceedings of the 5th International In Situ and On-site Bioremediation Symposium* (Bruce C. Alleman and Andrea Leeson, eds.), pp. 301-306, April 19-22, 1999, San Diego, CA.

Reed, P., Minsker, B., and Valocchi, A. J. (2000). "Cost-effective long-term groundwater monitoring design using a genetic algorithm and global mass interpolation." *Water Resources Research*, 36(12), 3731-3741.

Reed, P. M., Minsker, B. S., and Valocchi, A. J. (2001a). "Why optimize long term groundwater monitoring design? A multiobjective case study of Hill Air Force Base." *Proceedings of the ASCE World Water and Environmental Resources Congress, Orlando, FL.*

Reed, P., Minsker, B. S., and Goldberg, D. (2001b). "A multiobjective approach to cost effective long-term groundwater monitoring using an elitist nondominated sorted genetic algorithm with historical data." *Journal of Hydroinformatics*, 3(2), 71-89.

Reed, P. (2002). "Striking the balance: Long-Term groundwater monitoring design for multiple conflicting objectives," PhD Thesis, University of Illinois, 2002. http://cee.uiuc.edu/emsa/publications.html.

Reingruber, M. and Gregory, W. W. (1994). *The data modeling handbook: A best-practice approach to building quality data models*. Wiley, New York.

Revelle, C. S., Whitlach, E. E., and Wright, J. R. (1997). *Civil and Environmental Systems Engineering*, Prentice-Hall, Upper Saddle River, NJ.

Ridley, M., and D. MacQueen. (1995). "Cost-Effective Sampling of Groundwater Monitoring Wells: A Data Review & Well Frequency Evaluation." Hazardous

Materials Management Conference and Exhibition, April 4-6, San Jose, http://www-erd.llnl.gov/library/JC-118909.pdf.

Ridley, M. N., Johnson, V. M., and Tuckfield, R. C. (1995). "Cost-effective sampling of ground water monitoring wells," *HAZMACON 1995*, San Jose, CA, April 4-6, 1995. (Also available as *UCRL-JC-118909, Preprint*, Lawrence Livermore National Laboratory, Livermore, CA.).

Ridley, M., and MacQueen, D. (1995). "Cost-effective sampling of groundwater monitoring wells: A data review & well frequency evaluation." *Proceedings of the Hazardous Materials Management Conference and Exhibition, April 4-6, 1995, San Jose, California*, 14-21. Updated version at http://www-erd/library/.

Ridley, M N., and Johnson, V. (1996). "Cost-Effective sampling of ground water monitoring wells." In *Spectrum '96 Conference*, August 18-23, Seattle, WA. American Nuclear Society 1996. (Also available as *UCRL-JC-118909 Rev. 1, Preprint*, Lawrence Livermore National Laboratory, Livermore, CA.).

Rizzo, D. M., and Dougherty, D. E. (1996). "Design optimization for multiple management period groundwater remediation." *Water Resources Research*, 32 (8), 2549-2561.

Rizzo, D. M., Dougherty, D. E., and Yu, M. (2000). "An adaptive long-term monitoring and operations system (aLTMOs™) for optimization in environmental management," *Building Partnerships: Proceedings of 2000 ASCE Joint Conference on Water Resources Engineering and Water Resources Planning and Management, Minneapolis, MN* (ISBN 0-7844-0517-4), ed. R. H. Hotchkiss and M. Glade, American Society of Civil Engineers, Reston, VA.

Rizzo, D. M., Dougherty, D. E., and Yu, M. (2000). "An adaptive monitoring and operations system (aLTMOs™) for environmental management." In *Building Partnerships, Proc. 2000 ASCE Joint Conference on Water Resources Engineering and Water Resources Planning & Management*, R. H. Hotchkiss and M. Glade (eds.), American Society of Civil Engineers, Reston, VA, CD-ROM (ISBN 0-7844-0517-4).

Rizzo, D. M., Dougherty, D. E., and Yu, M. (2001). "Devising groundwater LTM strategies for different objectives using optimization," *Bridging the Gap, Proceedings of 2001 EWRI/ASCE World Water and Environmental Resources Congress, Orlando, FL* (ISBN 0-7844-0569-7), ed. D. Phelps and G. Sehlke, American Society of Civil Engineers, Reston, VA.

Rouhani, S. (1985). "Variance reduction analysis." *Water Resources Research*, 21(6), 837-846.

Rouhani, S. and Hall, T. J. (1988). "Geostatistical schemes for ground-water sampling." *Journal of Hydrology*, 81(1), 85-102.

Royle, J. A. (2000). "Optimal spatial network design for temporal trend estimation", in revision, *Journal of Statistical Planning and Inference*, 2000.

Scheibe, T. D., and Lettenmaier, D. P. (1989). "Risk-based selection of monitoring wells for assessing agricultural chemical contamination of ground water."*Ground Water Monitoring Review*, 9(4), 98-105.

Schrage, L. (1991). *LINDO: An Optimization Modeling System.* The Scientific Press, San Francisco, CA.

Seaborn, D. (1995). "Database management in GIS: Is your system a poor relation?" GIS Europe, 4(5):34-38, 1995.

Singh, S., and Harvey, A. T. (1999). "Process optimization of remedial systems." In USEPA, 2000, pp. 48-49, 1999.

Storck, P. J., Eheart, J. W., and Valocchi, A. J. (1997). "A method for the optimal location of monitoring wells for detection of ground-water contamination in three-dimensional aquifers." *Water Resources Research*, 33(9), 2081-2088.

Taylor, J. K. (1987). Quality assurance of chemical measurements, Lewis Publishers, Chelsea, Mich. http://www.frtr.gov/optimization/optimizc.html# LTMOptim.

Toregas, C., Revelle, C., and Bergman, L., 1971. The location of emergency service facilities. *Operations Research*, 19, 1363-1373.

Tucciarelli, T., and Pinder, G. (1991). "Optimal data acquisition strategy for the development of a transport model for groundwater remediation," *Water Resources Research*, 27(4), 577-588.

Tuckfield, C., and Ridley, M. (1999). "Workshop on Long-Term Monitoring Optimization," In U.S. Federal Remediation Technologies Roundtable, 1999.

Tuckfield, R.C., Shine, E.P., Hiersgesell, R.A., Denham, M.E., Beardsley, C., and Reboul, S. (2001). "Using geoscience and geostatistics to optimize ground-water monitoring networks at the Savannah River Site." *Proceedings of the Water and Environment Congress,* 1-17, ASCE Press, Reston, Virginia.

Ubeda, T. and Egenhofer, M. J. (2001). "Topological error correcting in GIS." advances in spatial databases, Fifth International Symposium on Large Spatial Databases, Berlin.

U.S. DoD. (1989). *Evaluation of DoD waste site ground-water pump-and-treat operations.* Office of the Inspector General, Report Number 98-090, 1989.

U. S. EPA (United States Environmental Protection Agency). (1986). *RCRA Ground-Water Monitoring Technical Enforcement Guidance Document.* Office of Solid Waste and Emergency Response, USEPA, Washington, D.C.

U.S. EPA. (1989). Statistical analysis of ground-water monitoring data at RCRA (Resource Conservation and Recovery Act) facilities, interim final guidance. U.S. EPA Report No. EPA/530-SW-89-026, Office of Solid Waste, USEPA, Washington, D.C.

U.S. EPA. (1992). Statistical analysis of ground-water monitoring data at RCRA Facilities: Addendum to interim final guidance. Office of Solid Waste, USEPA, Washington, D.C.

U. S. EPA. (1996). *Proposed guidelines for ecological risk assessment,* EPA/630/R-95/002B, Risk Assessment Forum, USEPA, Washington, DC.

U.S. EPA. (2000). *Subsurface remediation: Improving long-term monitoring and remedial systems performance.* Conference Proceedings June 8-11, St. Louis, MO. EPA/542/B-00/002. USEPA, Washington, DC, 2000.

USGS. (1998). "Arsenic in ground water of the Willamette Basin, Oregon. Data Quality: Sampling and analytical variability versus environmental variability." http://oregon.usgs.gov/pubs_dir/Online/Html/WRIR98-4205/.

U.S.G.S. (1998). "Data quality objectives and criteria for basic information, acceptable uncertainty, and quality-assurance and quality-control documentation." 98-394, http://ma.water.usgs.gov/fhwa/products/ofr98-394.pdf, U.S. Department of Transportation.

U.S.G.S. (2000). "Interagency field manual for the collection of water-quality data." http://water.usgs.gov/pubs/ofr/ofr00-213/manual_eng/collect.html.

U.S. Navy. (2001). "Ecological RISK ASSESSMENT PRocess," *Navy Guidance for Conducting Ecological Risk Assessments,* online at http://web.ead.anl.gov/ecorisk/process/index.cfm.

U.S. Navy. (2001). RAO-LTM web page http://erb.nfesc.navy.mil/support/work_grp/raoltm/main.htm (June 21, 2001).

Van Ee, J. J., Blume, L. J., and Starks, T. H. (1990). "A Rationale for the assessment of errors in the sampling of soils." EPA/600/4-90/013, EPA, Las Vegas.

Wagner, J. M., Shamir, U., and Nemati, H. R. (1992). "Groundwater quality management under uncertainty: Stochastic programming approaches and the value of information," *Water Resources Research*, 28(5), 1233-1246.

Wagner, B. (1995). "Sampling design methods for groundwater modeling under uncertainty." *Water Resources Research*, 31(10), 2581-2591.

Wilson, N. (1995). *Soil water and ground water sampling*, Lewis Publishers, Boca Raton, FL.

Yeh, T.-C. J. (1992). "Stochastic modeling of groundwater flow and solute transport in aquifers." *Hydrological Processes*, 6, 369-395.

Zheng, C. (1990). MT3D—A modular three-dimensional transport model for simulation of advection, dispersion, and chemical reactions of contaminants in ground-water systems. S. Papadopulos and Associates, Inc., Rockville, Maryland.

Zheng, C. (1990). MT3D: A modular three-dimensional transport model for simulation of advection, dispersion and chemical reactions of contaminants in groundwater systems. S.S. Papadopulous and Associates, Rockville, MD.

Index

IRDMIS *see* Installation Restoration Data Management Information System

Kalman filter methods 47–48, 60(table), 74–76
karst aquifers 35, 79
kriging 44, 45–46, 70

laboratories: checking data quality 19–20
landfills: case study 60(table), 62–63, 63(figure); hydrogeologic method 35–36; monitoring design 48
Lawrence Livermore National Laboratory (LLNL) 23–24, 37, 61(table)
LINDO 60(table), 62
linear regression methods 42
"living" performance assessment models 77–78
Location Set Covering Model 60(table), 62
long-term monitoring (LTM): definition 6–8; need for 5–6; *vs.* other forms of monitoring 7
Long-Term Monitoring Optimization (LTMO) ToolKit 68

Mann-Kendall test 37
MAROS *see* Monitoring and Remediation Optimization System
Massachusetts Military Reserve 60(table), 70–71
mathematical optimization methods 49–54; genetic algorithms 53–54; integer programming 51; simulated annealing 52–53, 52(equation)
matrices: Field Sampling and Analysis Technologies 20; soil 21; space-time covariance 47; water 21
maximum contaminant level (MCL) 39, 40
McClellan Air Force Base (Calif.) 60(table), 61(table), 67

methods 31–56; background well to compliance well comparisons 39–40; comparisons with fixed standards 40–41; by complexity level 32(table); concentration-based geostatistical 45–46; executive summary 2–3; genetic algorithms 53–54; geostatistical 42–46, 43(equation), 44(equation); hierarchical 48–49; hybrid hydrogeology-geostatistics 46; hydrogeologic 34–36, 36(figure); integer programming 51, 54–56, 55(equation), 56(equation); intra-well comparisons 41, 41(equation); Kalman filter 47–48, 60(table), 74–76; mathematical optimization 49–54; new 9–10; by objectives 33(table); probabilistic 46–49; rule-based 34–37; selecting 38–39, 38(figure); simulated annealing 52–53, 52(equation); statistical 37–46; time series data 54, 54(equation); trend detection 41–42; variance-based geostatistical 44–45
Metropolis algorithm 52
models: correcting 15; data 18; flow and transport 34–35, 47, 52, 60(table), 71–73; Location Set Covering Model 60(table), 62; MODFLOW 35, 62; MODFLOW-SURFACT 3-D simulation 71; MT3D 35, 62; performance assessment 77–78; predictive 13; site 15; transport 60(table), 74–76
MODFLOW (flow model) 35, 62
MODFLOW-SURFACT 3-D simulation model 71
monitored natural attenuation (MNA) 58, 59
Monitoring and Remediation Optimization System (MAROS) 37, 60(table), 67
MT3D (transport model) 35, 62